AF749214

Vehicle Technology and Automotive Engineering

Edited by Hasan Koten and Selçuk Sarıkoç

Published in London, United Kingdom

Vehicle Technology and Automotive Engineering
http://dx.doi.org/10.5772/intechopen.1006216
Edited by Hasan Koten and Selçuk Sarıkoç

Contributors
Abubakar Sadiq Adamu, Alejandro Yamaguchi Torres-Valdez, Alfredo Rios-Nogues, Andrei Rădulescu-Gamaleț, Andrew F. Burke, Anthony Witts, Arridina Susan Silitonga, Eduardo Macias-Avila, Georgiana Claudia Vasiliu, Hasan Koten, Islam Md Rizwanul Fattah, James Witts, Jesus M. Orona-Hinojos, Jingyuan Zhao, Jonathan Sherlock, Kubilay Guner, Lizbeth A. Huerta-Larrumbe, Marius Daniel Bontoș, Md Abul Kalam, Md Mofijur Rahman, Mohammad Rashed Mia, Morteza Mollajafari, Muhammad Ali Ijaz Malik, Nicholas Moulis, Nicolae Vasiliu, Saliha Özarslan, Salman Ebrahimi-Nejad, Selçuk Sarıkoç, Shuichi Matsumoto, T.M. Indra Mahlia, Tai Fei, Xiaoyu Sun, Yanjun Zhang, Zubairu Abubakar

© The Editor(s) and the Author(s) 2025

The rights of the editor(s) and the author(s) have been asserted in accordance with the Copyright, Designs and Patents Act 1988. All rights to the book as a whole are reserved by INTECHOPEN LIMITED. The book as a whole (compilation) cannot be reproduced, distributed or used for commercial or non-commercial purposes without INTECHOPEN LIMITED's written permission. Enquiries concerning the use of the book should be directed to INTECHOPEN LIMITED rights and permissions department (permissions@intechopen.com).

Violations are liable to prosecution under the governing Copyright Law.

Individual chapters of this publication are distributed under the terms of the Creative Commons Attribution 4.0 License which permits commercial use, distribution and reproduction of the individual chapters, provided the original author(s) and source publication are appropriately acknowledged. If so indicated, certain images may not be included under the Creative Commons license. In such cases users will need to obtain permission from the license holder to reproduce the material. More details and guidelines concerning content reuse and adaptation can be found at http://www.intechopen.com/copyright-policy.html.

Notice

Statements and opinions expressed in the chapters are these of the individual contributors and not necessarily those of the editors or publisher. No responsibility is accepted for the accuracy of information contained in the published chapters. The publisher assumes no responsibility for any damage or injury to persons or property arising out of the use of any materials, instructions, methods or ideas contained in the book.

First published in London, United Kingdom, 2025 by IntechOpen
IntechOpen is the global imprint of INTECHOPEN LIMITED, registered in England and Wales, registration number: 11086078, 167-169 Great Portland Street, London, W1W 5PF, United Kingdom

For EU product safety concerns: IN TECH d.o.o., Prolaz Marije Krucifikse Kozulić 3, 51000 Rijeka, Croatia, info@intechopen.com or visit our website at intechopen.com.

British Library Cataloguing-in-Publication Data
A catalogue record for this book is available from the British Library

Vehicle Technology and Automotive Engineering
Edited by Hasan Koten and Selçuk Sarıkoç
p. cm.
Print ISBN 978-1-83634-866-5
Online ISBN 978-1-83634-865-8
eBook (PDF) ISBN 978-1-83634-867-2

If disposing of this product, please recycle the paper responsibly.

IntechOpen
intechopen.com

Built by scientists, for scientists

Explore all IntechOpen books

Meet the editors

Prof. Hasan Koten worked on the first full geometry engine design project for the Scientific and Technological Research Council of Türkiye (TUBITAK). During his Ph.D., he worked as a visiting scholar at Ohio State University in the United States. Hasan Koten also worked as a consultant for the TUBITAK Domestic Electric Vehicle Design project in 2017 and as a post-doctoral researcher in the CEDP group at Brunel University in 2018. He published 150 articles. He was awarded by TUBITAK 2238. Additionally, he worked at CCRC, KAUST, and the University of Applied Sciences in Germany as a visiting professor. He has many patents in the field of energy systems. Prof. Hasan Koten continues to serve as the head of the Mechanical Engineering Department at Istanbul Medeniyet University in Türkiye. Currently, he is working on Türkiye's largest TUBITAK budgeted 2400HP 16 and 8-cylinder engine production project.

Dr. Selçuk Sarıkoç is an Associate Professor in the Department of Mechanical Engineering at Amasya University in Türkiye. He earned his Bachelor of Science degree in 2011 and his Ph.D. in 2019 from Erciyes University in Türkiye, specializing in energy systems. Dr. Sarıkoç's research primarily focuses on alternative fuels, such as biodiesel and hydrogen, and their applications in internal combustion engines, as well as analyzing the energy efficiency and environmental impacts of these fuels. He has published over 60 academic works, including articles in SCI(E)-indexed journals, and holds a nationally registered patent. Dr. Sarıkoç was supported by the TÜBİTAK 1512 BIGG Entrepreneurship Support Program. His project focuses on fuel enrichment and the development of advanced fuel additives for unmanned aerial vehicles (UAVs) in the defense industry. Thus, in 2023, he founded FSMS Engineering, a Technopark startup that aims to produce sustainable alternative aviation fuels via innovative fuel enrichment methods and the design of avionic systems for UAVs. Dr. Sarıkoç's work effectively bridges the gap between academic research and industrial applications, making significant contributions to both fields.

Contents

Preface

The automotive and vehicle industry is one of the most important transformations in the history of technology. Driven by the global imperatives of sustainability, digital innovation and intelligent automation, today's mobility solutions continue to develop rapidly in many areas, from powertrain efficiency and emission reduction to advanced driver assistance systems and autonomous vehicles. This book brings together contemporary research and engineering studies that reflect the multidisciplinary nature of today's automotive and car industry. It offers various topics, including experimental studies, computational studies, artificial intelligence applications and sustainable technologies. In this way, it encompasses studies that emerge as a whole from interdisciplinary research.

The common goal of the book is to integrate intelligent systems into traditional engineering frameworks. Applying machine learning, data compression, and digital twin technologies also exemplifies how data-driven methodologies reshape design, diagnostics and control processes. Meanwhile, research on combustion behavior, emission reduction and alternative fuel additives highlights the continuing importance of classical sciences, especially as the industry transitions to cleaner energy sources. In addition, the book reflects a global perspective, with contributions that address a range of challenges and opportunities, from the electrification of transportation systems to skills development for future job markets in electromobility. This diversity not only enriches the technical content but also provides a broader understanding of how innovation is being implemented in different parts of the world.

This book aims to be a valuable resource for researchers, practitioners and experts interested in the evolving landscape of automotive and vehicle engineering. It fosters a deeper understanding of both fundamental principles and state-of-the-art developments while also encouraging interdisciplinary collaboration and forward-looking design.

Hasan Koten
Mechanical Engineering Department of Istanbul Medeniyet University,
Istanbul, Türkiye

Selçuk Sarıkoç
Faculty of Engineering,
Department of Mechanical Engineering,
Amasya University,
Amasya, Türkiye

Chapter 1

Droplet Evaporation and Combustion Characteristics of Fuels: A Review

Saliha Özarslan and Selçuk Sarıkoç

Abstract

The evaporation of fuel droplets represents one of the fundamental stages in the combustion process. Liquid fuel injected into the combustion chamber undergoes evaporation in droplet form, followed by ignition and subsequent combustion. Understanding the droplet evaporation and combustion characteristics of different fuel types provides a foundation for enhancing combustion efficiency and improving emission control. This study presents a review of recent research on the droplet evaporation and combustion characteristics of a wide range of fuel droplets such as diesel, gasoline, nanofluid fuels, and biofuels. Experimental and numerical studies were investigated to evaluate the effects of various combustion conditions on the droplet evaporation of fuels. The focus of the study is to examine how parameters such as ambient temperature, pressure, droplet size influence droplet evaporation and combustion behavior. By analyzing the droplet evaporation and combustion properties of fuels, it is possible to increase combustion efficiency and advance combustion technologies. This review aims to contribute to the filling of the gap in the field and provide knowledge for future research studies.

Keywords: droplet evaporation, combustion characteristics, fuels, ambient conditions, droplet size

1. Introduction

Combustion is an important process used in many different fields, including heating, energy production, transport, petroleum refining, power generation, and manufacturing industries. The fact that it has a wide usage area shows that a significant proportion of energy use worldwide belongs to the combustion process. Today, as in many areas where energy consumption is in question, fossil fuels are mostly used in the combustion process [1]. However, with the increase in population and industrialization, energy use is increasing and non-renewable fossil resources are rapidly depleting. Considering that fossil fuels account for about 85% of total energy consumption, it is clear that urgent solutions are needed for energy use. Another concern is the damage caused by fossil fuels to the environment and human health [2]. Fossil fuel combustion releases greenhouse gases that contribute to air pollution and climate change. In recent decades, the generation and combustion of fossil fuels for energy and transport has released more than 120 million tons of methane and billions of tons of carbon dioxide into the atmosphere each year. Another consequence of fossil fuel emissions is the increase in the temperature of the earth's surface, also known as global warming [3].

In this context, in order to address the threat of fossil fuel depletion as well as the harm caused by them, researchers are still looking for alternative fuels that can take their place. Many wastes and biomass are seen as viable resources for the production of chemicals, fuels, and energy. Fuels such as biodiesel, bio-oil, bio-ethanol, biogas and bio-hydrogen produced from biomass are strong candidates to replace fossil fuels as they emit less carbon than conventional fuels and are renewable [4, 5]. For example, the reflection of the search for alternative fuels in practice can be seen in electric, hybrid-electric, and alternative fuel-burning vehicles. These vehicles, which are under development or whose production and use have already started, are promising in terms of reducing the use of fossil fuels [6]. However, the complete replacement of fossil fuel systems and the widespread use of renewable and environmentally friendly fuels is a process that requires time. The development of nuclear energy, energy efficiency enhancements, energy conservation, carbon taxes, a more equitable distribution of energy use per capita and human welfare, cap and trade systems, carbon capture, utilization, and storage, and the development of renewable energy are the eight pathways that will be necessary to achieve a full transition. By 2050, this would allow for the substitution of renewable energy for fossil fuels; however, it would necessitate the vigorous execution of all eight pathways, substantial lifestyle adjustments in industrialized nations, and close international cooperation. While the search for, development, and implementation of alternative energy continues, in order to save time and create better conditions in this process, improvement of existing systems, efficient use of energy, and reduction of harmful effects are also issues that need to be addressed [7].

This is clearly visible in the transport sector, where the combustion process is used on a large scale. Today, 99.8% of transportation uses internal combustion engines, and 95% of fuel needs are met by petroleum-based liquid fuels. Due to the difficulties in switching to alternative fuels, it is expected that by 2040, 85–90% of conventional liquid fuels will still be used. Therefore, it becomes essential to improve the combustion process, increase the efficiency of the system, reduce exhaust emissions, and use energy efficiently [8]. The first step is to understand the combustion process and the behavior of fuels and to know the effect of combustion parameters and ambient conditions on the process. It is valuable to bring together the studies carried out by a large number of researchers using different fuels, various environmental conditions, and different methods and to present them to the reader's interest in order to accelerate the development of the subject studied and to increase awareness.

Studies about the combustion properties of fuels and the evaporation process have been compiled and examined by researchers from time to time. For example, Mehrizi et al. [9] concentrated on how additive nanoparticles affected the mechanism and evaporation behavior of fuel droplets. It was shown how many factors, including the kind, concentration, size, and ambient temperature of the nanoparticles, affect the evaporation properties. Aggarwal [10] has compiled experimental, theoretical, and computational studies on the ignition of single droplets in a stagnant and high-pressure environment by spray ignition. Multicomponent droplet ignition and convective effects on droplet ignition are also among the topics investigated. Zhu et al. [11] investigated in detail the different applications of water injection in internal combustion engines and water evaporation stages.

In another review, the physicochemical parameters of biodiesel-based fuels including CeO_2 nanoparticles, as well as the combustion behavior, performance, and emission properties of diesel engines running on biodiesel containing CeO_2 nanoparticles, were thoroughly examined by Hoang [12]. Studies about the fundamental combustion properties of ethanol and biodiesel, such as counter-flow diffusion flame, single-droplet combustion, co-flow diffusion flame, spray combustion, and wick-induced flame, as well as the impact of additives on pollutant generation, were compiled by Lee et al. [13]. Téré et al. [14] studied the numerical modeling of the natural convection evaporation of a liquid fuel droplet, investigating the advection, azimuth angle, and thickness of the vapor phase of the droplet during the evaporation process. Sazhin [15] reviewed the modeling studies of fuel droplet heating and evaporation between 2014 and 2017 and identified unsolved problems. Wang et al. [16] reviewed the experimental studies on the evaporation properties of different fuel droplets.

The studies given above mostly serve a narrower scope, such as the behavior of a specific fuel type and results for specific conditions. This paper is a comprehensive review of recent experimental and numerical studies on the combustion characteristics of fuels and droplet evaporation. For various fuels and under different conditions, the effects of ambient conditions, such as temperature and pressure, and fuel properties, such as droplet size, on the droplet evaporation and combustion characteristics of fuels were investigated. Knowing how parameters such as ambient temperature, ambient pressure and droplet size affect the combustion process under which conditions and for which fuels will enable the improvement of combustion efficiency and more efficient use of fuels. For this purpose, it is believed that the current review will contribute to the development of present practices and the guidance of new studies.

2. Materials and methods

2.1 Motivation and outline

Droplet evaporation of fuels is an important process in the combustion process. The combustion stage starts with droplet evaporation, followed by ignition and combustion. The droplet vaporization process is influenced by a wide range of parameters for each fuel type. In this study, the effects of temperature, pressure, and droplet size, which are important combustion parameters for different fuel types, on droplet evaporation and combustion of fuels were investigated. In the following part of the study, information about the methods of the studies on droplet evaporation and combustion properties of fuels are given and exemplified. In the third part, which is the main part, the investigations are presented under the headings of the effects of ambient conditions and fuel properties. In each section, firstly, experimental and then numerical studies on temperature, pressure, and droplet size are given. Finally, in the 'Conclusions' section, the effects of the analyzed parameters are summarized, and conclusions are presented. It is thought that the investigation of different fuel types and operating conditions will provide an opportunity to improve the literature resources and fill the existing gaps. It is also aimed to shed light on related studies such as improving the combustion process and increasing fuel efficiency.

2.2 Methods of studies on droplet evaporation and combustion characteristics

2.2.1 Experimental methods

Experimental methods about droplet evaporation and combustion properties of fuels are the first sources to obtain information about combustion conditions in real applications. In addition to improving combustion properties and developing combustion models, experimental methods are frequently used to validate studies. Therefore, as a result of the research, it is seen that the studies conducted in this field recently are mostly experimental. There are generally three experimental methods used to study the droplet evaporation of fuels. These are suspension, levitation, and flying droplets. Suspension is the process of suspending fuel droplets by means of thermocouples, quartz wires, silica fiber or other fibers. Levitation refers to the lifting of fuel droplets into the air by buoyancy, magnetic or acoustic forces. The flying droplet method involves the free-falling motion of fuel droplets [16]. The majority of the reviewed studies use the suspension method. An example experimental setup scheme is shown in **Figure 1**. In this example, the droplet is suspended in a K-type thermocouple. After that, an electric pushing device exposes it to a high-temperature chamber, where evaporation takes place [17].

2.2.2 Numerical modeling

Numerical studies about droplet evaporation and combustion properties of fuels are useful resources for predicting how droplets will actually react under what ambient conditions. The predictions obtained using comprehensive numerical models allow a comprehensive assessment of parameters such as flow and temperature of the evaporation and combustion stages [18]. Numerical studies about droplet evaporation

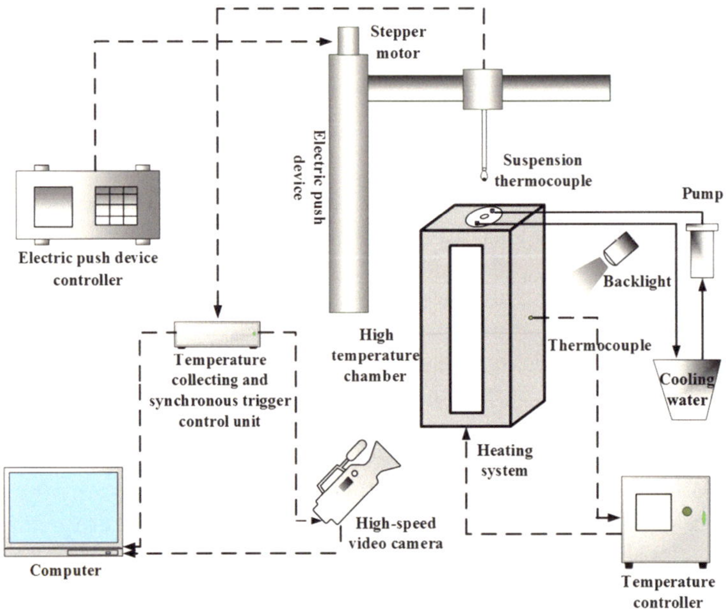

Figure 1.
Diagrammatic illustration of the experimental setup for single-droplet evaporation. Reprinted from Ref. [17].

and combustion are fewer than experimental studies in the literature. In numerical studies, different technologies and applications can be used, and studies can be analyzed by making certain assumptions. The accuracy of the modeling and simulations is checked by comparing them with other studies or experiments.

For example, in a study, a cubic simulation box was prepared, and a single suspended n-dodecane droplet in nitrogen was analyzed. The simulations were performed on a Cray XC30 supercomputer with the help of a micro-canonical ensemble (NVE) [19]. In another study, the following assumptions were made:

- The droplet starts off as a single chemical component and is constantly stationary.
- An equilibrium phase is present at the gas-liquid interface.
- In an inert atmosphere, the droplet evaporates.
- Because of the low Weber number, which fluctuates below roughly 2, the impact of droplet deformation on the droplet temperature and mass transfer can be disregarded.
- It is possible to ignore the effects of gravity, energy flow because of mass concentration (Dufour), and mass diffusion because of temperature (Soret) [20].

3. Effect of parameters droplet evaporation and combustion characteristics

3.1 Effect of ambient conditions

The conditions of the environment where combustion takes place directly affect the combustion process. Parameters such as temperature, pressure, and turbulence are the environmental conditions that affect combustion for all types of fuel. In this section, the effects of temperature and pressure on droplet evaporation for different types of fuel are compiled and presented.

3.1.1 Temperature

The heat transfer between the droplet and its surroundings is determined by the surrounding temperature. A higher ambient temperature causes droplets to absorb more heat, which speeds up droplet evaporation and reduces droplet longevity [16]. Ma et al. [21] experimentally investigated the evaporation characteristics of acetone-butanol-ethanol (ABE) and diesel mixture droplets at various ambient temperatures (423–823 K) using the suspension method. During the droplet evaporation process in the experiments, the droplet was heated by conduction and thermal radiation. As a result, even though the evaporation process of ABE droplets was essentially the same as that of diesel, it was discovered that ABE droplets always evaporated more quickly than diesel at various medium temperatures. However, as the ambient temperature rises, the droplet lifespan differential between ABE and diesel diminishes. Diesel-ABE blends also evaporate more quickly than diesel. Both ABE and the evaporation properties of diesel droplets follow the classical d^2 law. The ambient temperature affects the ABE-diesel mixes' droplet evaporation properties. The normalized square drop diameter changes for diesel and ABE fuels at different ambient temperatures are displayed in **Figure 2**. For all fuels, it is observed that the droplet life reduces as the ambient temperature rises.

Javed et al. [22] experimentally investigated the evaporation properties of n-heptane droplets containing different ratios of aluminum nanoparticles (0.5, 2.5, and 5%) suspended on silicon carbide fibers at different medium temperatures (100–600°C). The investigations showed that at high temperatures, heptane-based nanofluid fuel droplets obey the classical d^2 law and evaporate similarly to liquid fuel droplets. It is evident that the evaporation rates of fuel droplets based on heptane are highly influenced by the surrounding temperature. The change in the lifetime of pure heptane droplets depending on the temperature is given in **Figure 3**. As can be understood from the graph, the lifetime of pure heptane droplets decreases with the increase in ambient temperature. In stabilized heptane droplets, three-stage evaporation is observed. However, it was stated in the study that the evaporation behavior of stabilized heptane droplets also depends on medium temperatures. Javed et al. [23] also studied the auto-ignition and combustion properties of a single heptane (n-C_7H_{16}) droplet containing different concentrations of aluminum nanoparticles mounted on silicon carbide fiber by subjecting it to a rapid temperature increase from room temperature in the range of 600–850°C. At all temperatures (600–800°C), n-Al/heptane droplets exhibited destructive combustion behavior and did not obey the classical d^2 law, unlike pure heptane droplets. The gasification rate of 2.5 and 5.0% n-Al/heptane droplets was similar to or slightly lower than that of pure heptane droplets at relatively low temperatures (650–700°C). Regardless of the NP concentration, the gasification rate of nanofluid droplets was significantly faster than that of pure heptane droplets at higher temperatures (750–850°C). Normal gravity and atmospheric pressure were used for the studies. According to the previous study, the time-dependent change of the square of the diameter of pure heptane droplets in higher temperature environments is given in **Figure 4**. Accordingly, a heptane droplet with a diameter of 1 mm did not ignite at 550°C. However, ignition occurred at temperatures between 600 and 850°C. Then, the droplet expanded and burst, forming a smaller droplet. The droplet diameter then decreased according to the d^2 law.

Javed et al. [24] conducted an experimental investigation on the evaporation characteristics of kerosene droplets containing dilute concentrations (0.1, 0.5, and 1.0%) of ligand-protected Al nanoparticles suspended on silicon carbide fibers at

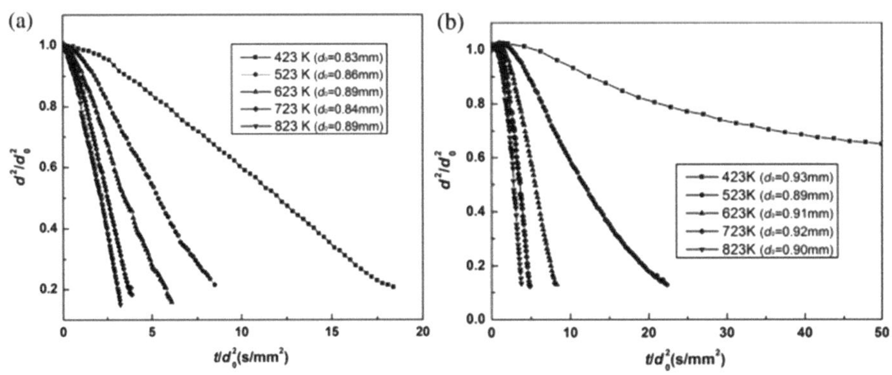

Figure 2.
Variations of droplet diameter for ABE (a) and diesel (b) at different medium temperatures. Reprinted from Ref. [21].

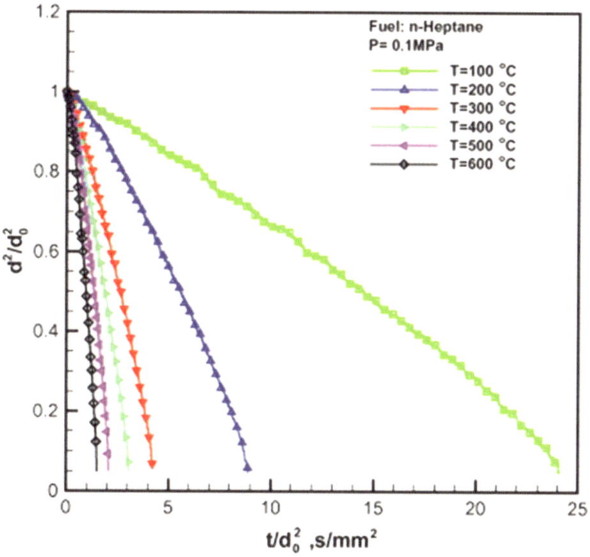

Figure 3.
Variation of pure heptane droplet diameter at different medium temperatures (100-600°C). Reprinted from Ref. [22].

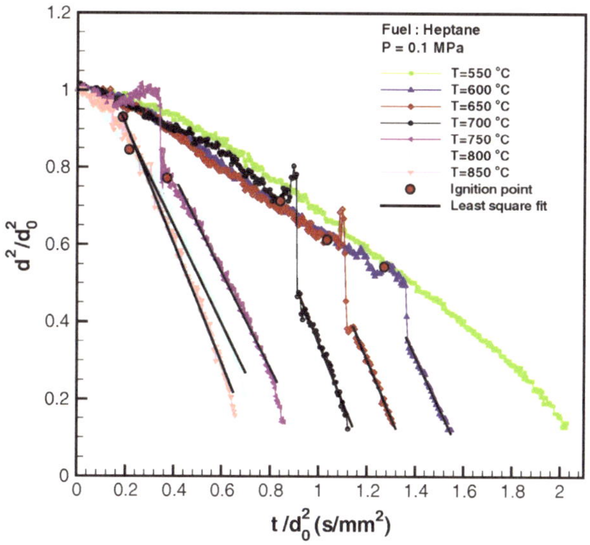

Figure 4.
Variation of pure heptane droplet diameter at different medium temperatures (550-850°C). Reprinted from Ref. [23].

varying medium temperatures ranging from 400 to 800°C. Kerosene-based nanofluid fuel droplets containing dilute aluminum nanoparticle concentrations followed the d^2 law at temperatures between 400 and 600°C. As a result, microexplosions were observed in kerosene-based nanofluid fuel droplets at relatively high temperatures (700–800°C). The evaporation rate of all aluminum nanoparticle suspensions consistently exceeded that of stable and pure kerosene droplets within the temperature range of 400–800°C, with a minor increase observed at comparatively low temperatures. However, the melting of aluminum nanoparticles resulted in a notable increase in evaporation at relatively high temperatures as 700–800°C. Javed et al. [25] experimentally investigated the auto-ignition and combustion properties of kerosene-based nanofluid droplets of dilute nanoparticle (NP) concentrations at high temperatures. The study was carried out by suspending an isolated kerosene droplet suspended on silicon carbide (SiC) fiber and containing different weight percentages (0.1, 0.5, or 1.0%) of aluminum (Al) NP at atmospheric pressure (0.1 MPa) and abruptly exposed to temperatures between 400 and 800°C. As a result of the study, unlike pure and stabilized kerosene droplets, n-Al/kerosene droplets showed destructive combustion behavior at all combustion temperatures and did not obey the classical d^2 law. However, dilute loading of nanoparticles reduced the intensity of these disturbances.

Han et al. [17] performed an experimental investigation to examine the evaporation properties of droplets in ethanol diesel blend fuel. Accordingly, they found that the DE10 (85% diesel +10% ethanol +5% biodiesel) droplet had a longer fluctuation evaporation stage than diesel (D100), and its lifespan was shorter at 623 K. They stated that the eruption and microexplosion were strengthened at 723 K. Yu et al. [26] investigated the evaporation characteristics of a multicomponent mixture consisting of ethanol, hydrogen, biodiesel, and diesel. They found that the evaporation of binary and ternary fuel droplets shortened, and the evaporation rate increased with increasing ambient temperature for mixtures containing multiple components with different volatilities in different temperatures (773, 873, and 973 K). They also stated that the increase in temperature contributed to the fluctuating evaporation of droplets and shortened the evaporation time. When they examined the droplet evaporation characteristics of HB10 (10% partially hydrogenated cottonseed methyl ester +90% diesel) and HB10E10 (10% partially hydrogenated cottonseed methyl ester +10% ethanol +80% diesel) fuels under various temperature conditions, they found that since the fuels consisted of multiple components with different volatilities, the lighter components expanded and evaporated first, while the heavier components went through certain stages.

Hashimoto et al. [27] experimentally studied the evaporation properties of a palm methyl ester droplet at different medium temperatures. The results were compared with diesel and n-hexadecane. Accordingly, it was determined that the initial heating period decreased, and the droplet life shortened with the increase in medium temperature for all fuels. The droplet life of palm methyl ester (PME) was found to be longer than that of diesel fuel and n-hexadecane. For a fuel with a low boiling point, the period during which the volumetric evaporation rate is lower than the thermal expansion rate is short. The boiling points of the constituents of diesel fuel are lower than the boiling points of the constituents of palm methyl ester. Therefore, at the same medium temperature, the initial heating period of diesel fuel and n-hexadecane droplets is shorter than that of PME droplets. During the evaporation stage of the droplets, it was observed that at a medium temperature of 473 K (**Figure 5**), the evaporation rate of the PME droplet is very low when D^2/D_0^2 is below 0.05. As seen in **Figure 5(a)**, the D^2/D_0^2 curve for diesel fuel is significantly different from other fuels. The authors stated that this is due to the decrease in the coefficient of instantaneous evaporation during the main evaporation stage.

Tao et al. [28] observed the combustion and evaporation properties of lubricating oil and gasoline droplets at ambient temperatures ranging from 673 to 873 K in their experimental study using suspension technology. Lubricating oil (LO) used lubricating oil (ULO) and a mixture of lubricating oil and gasoline (LO/gasoline and ULO/gasoline) were used in the study. When the $(d/d0)^2$ changes of LO and ULO droplets at

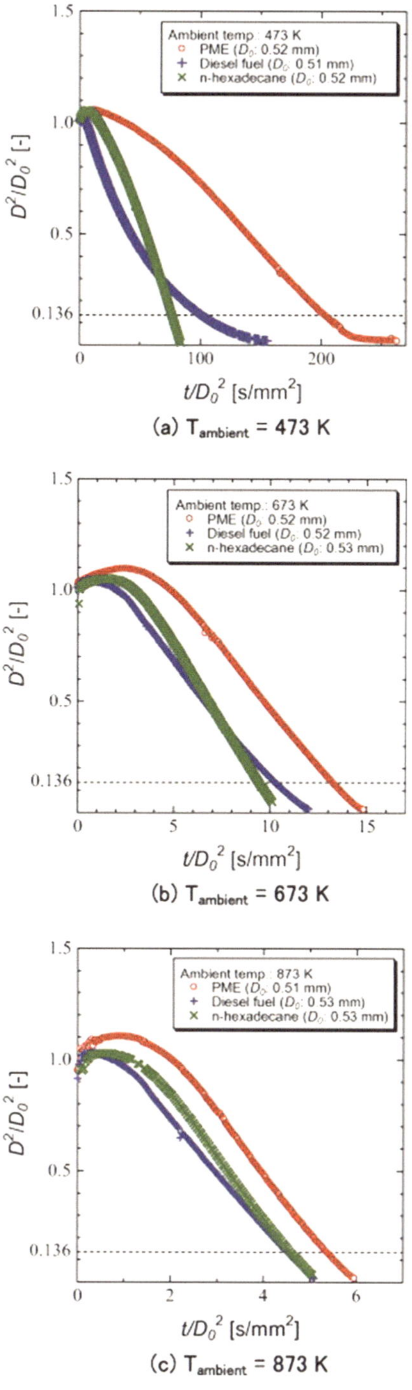

Figure 5.
Normalized D^2 histories of palm methyl ester, diesel fuel, and n-hexadecane droplets at temperatures of 473 K (a), 673 K (b), and 873 K (c). Reprinted from Ref. [27].

723 and 773 K in **Figure 6** are examined, it is seen that the fluctuation behavior of ULO occurs earlier than LO. The authors explained that this is because ULO contains metal,

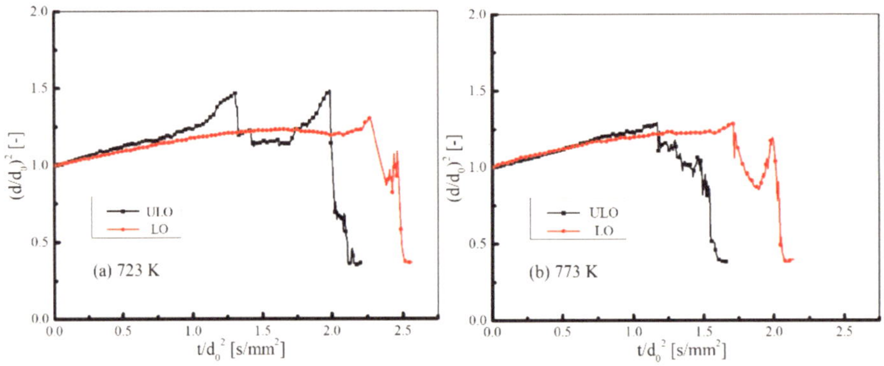

Figure 6.
Comparison of lubricating oil and used lubricating oil droplets $(d/do)^2$ at different ambient temperatures. Reprinted from Ref. [28].

which allows bubble nucleation to occur easily, and water with a low boiling point. According to the data obtained from the overall study, auto-ignition cannot occur below 673 K; however, auto-ignition occurs between 723 and 873 K. An auto-ignition delay interval, a combustion phase, and an extinction period make up the combustion process. As the medium temperature rises, the auto-ignition delay falls. ULO has a shorter lifetime and auto-ignition delay than LO at the same ambient temperature.

Xiao et al. [19] studied the evaporation stages of an n-dodecane droplet by molecular dynamics simulation. Suspended fuel droplets of various diameters were evaporated in a nitrogen environment, and the effects of medium pressure and temperature were observed. Within the scope of the study, it was determined that evaporation properties are affected by ambient conditions and droplet size parameters. Accordingly, as the medium temperature and/or pressure rise, the droplet lifespan decreases. It was also found that droplet lifetime decreased monotonically with increasing medium pressure at all medium temperatures. **Table 1** shows the studies whose methods and results are examined in this section.

3.1.2 Pressure

Droplet life increases as medium pressure increases. However, the effect of pressure under different conditions in recent studies in the literature has been examined in this section. Verwey and Birouk [29] experimentally studied the effect of droplet size and turbulence on the evaporation rate of suspended monocomponent alkane droplets (n-heptane, n-decane) at pressures up to 10 bar. As a result, the profiles of larger droplets deviate from the d^2 behavior at later stages, while smaller droplets evaporate rapidly and show a linear d^2 profile everywhere. **Figure 7** shows the evaporation lifetime of large (a) and small (b) decane droplets at different pressures (1, 5, and 10 bar). As can be seen from the graphs, droplet life increases as pressure increases. Birouk and Fabbro [30] experimentally studied n-heptane droplet evaporation at a medium temperature under high pressure (up to 21 atm) and turbulent conditions. Experiments have shown that as the ambient pressure rises, the droplet evaporation rate falls. It has also been found that at high pressure, turbulence intensity becomes much more effective in increasing droplet gasification. In another study where similar results were obtained, Birouk [31] experimentally studied the evaporation process of n-heptane and n-decane droplets in environments with pressure up

Reference	Operating range (Temperature)	Fuel	Method
Ma et al. [21]	423–823 K	Acetone-butanol-ethanol and diesel blends	Experimental/ suspension
Javed et al. [22]	100–600°C	n-Heptane containing different ratios of aluminum nanoparticles	Experimental/ suspension
Javed et al. [23]	600–850°C	n-Heptane containing different concentrations of aluminum nanoparticles	Experimental/ suspension
Javed et al. [24]	400–800°C	Kerosene containing dilute concentrations of ligand-protected aluminum nanoparticles	Experimental/ suspension
Javed et al. [25]	400–800°C	Kerosene containing dilute aluminum nanoparticles	Experimental/ suspension
Han et al. [17]	623 K 723 K	Ethanol diesel blend	Experimental/ suspension
Yu et al. [26]	773, 873, 973 K	A multicomponent mixture consisting of ethanol, hydrogen, biodiesel, and diesel	Experimental/ suspension
Hashimoto et al. [27]	473, 673, 873 K	Palm methyl ester, diesel fuel, and n-hexadecane	Experimental/ suspension
Tao et al. [28]	673–873 K	Lubricating oil, used lubricating oil, and the mixture of lubricating oil and gasoline	Experimental/ suspension
Xiao et al. [19]	Ranging from subcritical to supercritical	n-Dodecane into nitrogen environment	Numerical/ suspension

Table 1.
Studies analyzed in the temperature section.

to 16 bar, turbulent intensity up to 3.10 m/s and temperature up to 150°C. In all the ambient conditions examined in the study, the hydrocarbon droplet obeyed the d^2-law all of its life. As a result, it was observed that the droplet life increased as the pressure increased, and the droplet life decreased as the medium temperature increased. It was also reported that turbulence became more effective as the ambient pressure increased, but this effect decreased as the medium temperature increased.

Ma et al. [32] investigated the droplet evaporation and combustion properties in sub- and supercritical pressure environments in their experimental studies using hydrocarbon as fuel. Normal gravity was used for the experiments. According to the results, the ratio of droplet evaporation time to droplet burning time changed slightly in subcritical pressure environments. This ratio decreased rapidly in supercritical pressure environments. Therefore, droplet evaporation was completed earlier in supercritical pressure environments than in subcritical pressure environments. Birouk and Toth [33], experimentally studied the turbulent combustion of hydrocarbon (n-heptane and n-decane) droplets at high pressure. In their study, they determined that the droplet burning rate is insensitive to the turbulence level at atmospheric ambient pressure and becomes sensitive only at high ambient pressure. They observed that the droplet burning rate decreases linearly as the turbulence intensity increases. When the effect of ambient pressure on the droplet burning velocity at two typical turbulence intensities (0.13 and 0.62 m/s) was examined, it was determined that in the low turbulence

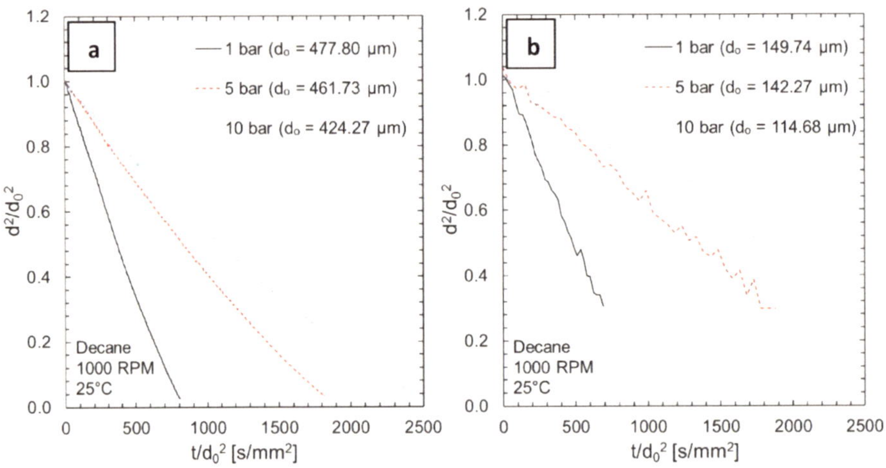

Figure 7.
Variation of diameter of large (a) and small (b) decane droplets at different pressures. Reprinted from Ref. [29].

range, the droplet burning velocity increases with increasing medium pressure, but the effect of pressure decreases as the turbulence intensity increases.

Birouk et al. [20] developed a three-dimensional numerical model to investigate droplet evaporation under high temperature and pressure conditions in a turbulent medium. In their study, they used 0.5–10 MPa free stream pressure, 324–502 K temperature, and 0–60% turbulence intensity. As a result, droplet life increases and, thus, the evaporation rate reduces as the pressure rises for the temperature range examined in the study. Nevertheless, it was found that the effect of pressure disappears when the medium temperature approaches the critical temperature of the n-heptane droplet. Kitano et al. [34] investigated the effects of medium pressure (0.1–2.0 MPa), initial gas temperature (1000–2000 K), and combustion reaction on droplet evaporation using three-dimensional numerical modeling. They used a non-equilibrium Langmuir–Knudsen evaporation model. According to the data obtained from the study, at a low initial gas temperature (1000 K), the droplet lifespan increases as medium pressure rises. However, it falls at high starting gas temperatures (1500 and 2000 K).

Kim and Sung [35] studied the effect of medium pressure on the evaporation of n-heptane droplets and spray using the high-pressure evaporation model. For spray

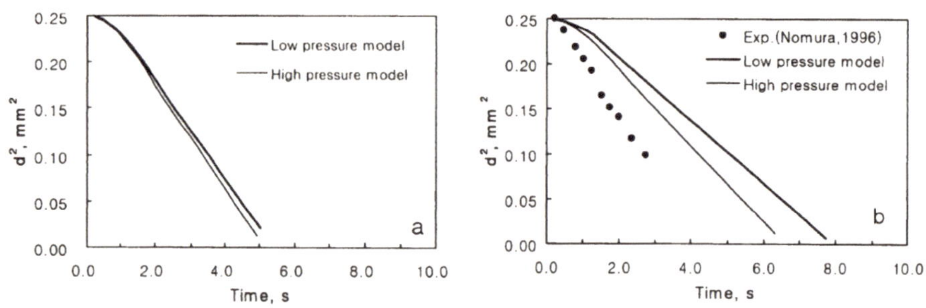

Figure 8.
Change of droplet's diameter square with respect to time at 20 atm (a) and 50 atm (b) pressure environments and 453 K ambient temperature (spray evaporation). Reprinted from Ref. [35].

evaporation, the evaporation process was investigated by injecting n-heptane droplets into gaseous nitrogen at high pressure. The initial diameter of the droplet at an initial temperature of 293 K is 0.5 mm. The history of the square diameter of the droplet at 20 and 50 atm ambient pressures is shown in **Figure 8**. The difference between the two models is more evident in **Figure 8b**. Based on this, the authors concluded that real/non-ideal gases should be investigated at high pressures. The predicted results were compared with an experimental study. It was stated that the deviation occurred because the energy transferred to the droplet was used to raise its temperature equally. As a result, it was found that the evaporation of a spray increased at higher ambient pressures. In the same study, it was found that in the evaporation of a single droplet, the life of the droplet increased with pressure when the ambient temperature was low and decreased with pressure when the ambient temperature was high. According to the data obtained from the study, it was concluded that the life of the droplet was longer at high pressures and low ambient temperature and shorter at high ambient temperature. The authors stated that this is due to the increase in temperature of the droplet in the final stages of evaporation at high pressure.

In their work, Aggarwal and Mongia [36] investigated the evaporation process of a binary component fuel droplet using various evaporation models such as diffusion limit and infinite-diffusion models and presented the results by comparing them with the evaporation characteristics of the equivalent single component droplet. It is observed that the simulation of the binary fuel with a surrogate fuel becomes progressively better as the pressure increases due to the increased droplet warm-up time and the decreasing volatility difference at higher pressures. In the studies conducted at 1 and 15 atm pressures, it was determined that the droplet lifetime for both binary and single component fuel droplets did not change significantly with ambient pressure. Nevertheless, the droplet warm-up time rose from around 20–40% of the droplet lifetime as the pressure was raised from 1 to 15 atm. They stated that the increase in

Reference	Operating range (pressure)	Fuel	Method
Verwey and Birouk [29]	Up to 10 bar	Monocomponent alkane (n-heptane and n-decane)	Experimental/suspension
Birouk and Fabbro [30]	Up to 21 atm	n-Heptane	Experimental/suspension
Birouk [31]	Up to 16 bar	Hydrocarbon	Experimental/suspension
Ma et al. [32]	Sub- and supercritical	Hydrocarbon	Experimental/suspension
Birouk and Toth [33]	1–11 bar	Hydrocarbon	Experimental/suspension
Birouk et al. [20]	0.5–10 MPa	n-Heptane	Numerical/freestream turbulence
Kitano et al. [34]	0.1–2.0 MPa	n-Decane and n-heptane	Numerical/a single and multiple
Kim and Sung [35]	5, 20, 40 bar 20, 50 atm	n-Heptane	Numerical/a single and a spray
Aggarwal and Mongia [36]	1–15 atm	Binary component gas turbine fuels	Numerical/infinite-diffusion and diffusion limit models

Table 2.
Studies analyzed in the pressure section.

droplet warm-up time was due to the increase in fuel boiling temperature with pressure. The studies reviewed in this section are given in **Table 2**.

3.2 Effect of fuel properties

Fuel properties are also factors that significantly affect the combustion process, such as ambient conditions. Knowing the effects of fuel type-specific parameters such as fuel type, number of components, concentration, and droplet size on droplet evaporation will provide an advantage in improving combustion. In this section, the effects of droplet size, one of the fuel properties, on evaporation and combustion for various fuel types are compiled and presented.

3.2.1 Droplet size

Fuel droplets smaller and larger than the optimum size cause problems such as non-ignition and ignition delay. As the diameter increases, droplet heating and vapor transport times increase, and ignition delay occurs. A decrease in diameter to a level less than optimum also causes ignition delay, and the droplet becomes unignitable [10]. In this section, examples from recent studies about the effects of droplet size on evaporation and combustion under various combustion conditions for fuel types will be presented.

Ghamari and Ratner [37] investigated the combustion properties of carbon-based nanoparticles and colloidal droplets using jet fuel. They studied the combustion

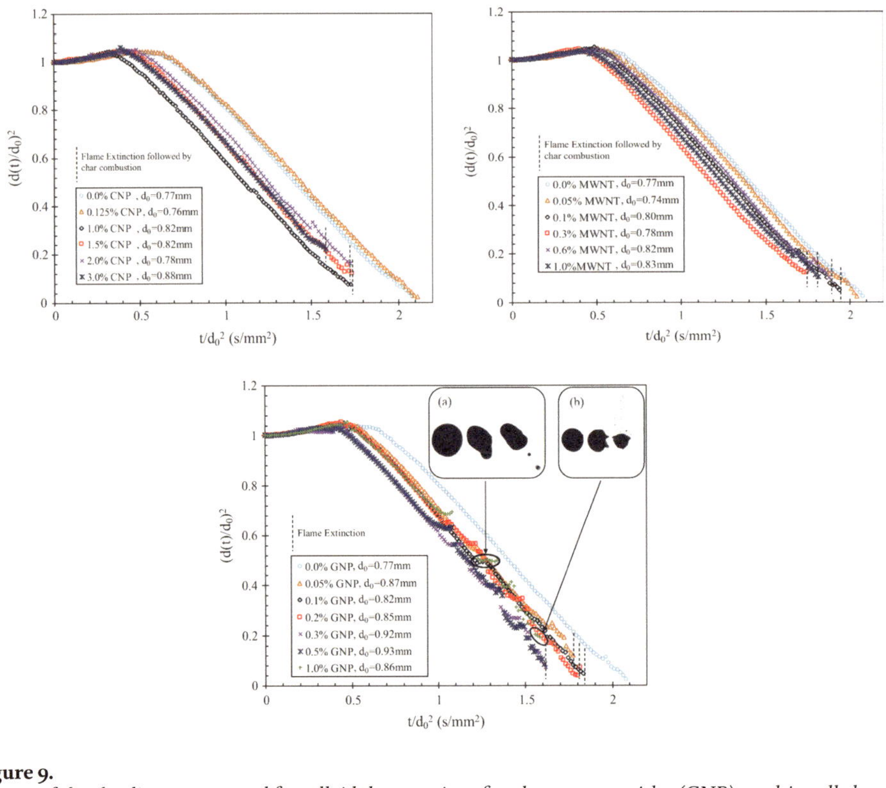

Figure 9.
History of droplet diameter squared for colloidal suspension of carbon nanoparticles (CNP), multi-walled carbon nanotubes (MWNT), and graphene nanoplatelets (GNP) in jet-A fuel (a. drop swelling and weak microexplosion, b. ejection of nanoparticles for 1.0% GNP in jet fuel droplet). Reprinted from Ref. [37].

properties of colloidal suspensions and sub-millimeter-sized droplets of various types of carbon-based nanoparticles by adopting the method of adding micro- and nano-sized particles to liquid fuels to improve their combustion properties. For this purpose, carbon nanoparticles, nanoplatelets, and nanotubes were added to jet fuel with sizes in the range of 2–100 nm and at various mass concentrations. **Figure 9** shows the percentage of additives and droplet diameters. Accordingly, it can be seen that the development of the droplet diameter squared in all colloidal suspensions follows the classical d^2 combustion law. As a result, a 7–10% higher combustion rate was obtained for all carbon additives, and it was determined that the size structure of the nanoparticles greatly affects the combustion rate. As with multicomponent fuels, preferential combustion is dominant, and the highly volatile basic fuel burns first. The reason for the few disturbances seen during combustion is that solid particles near the droplet surface act as nucleation points, causing weak microexplosions.

Nanlohy et al. [38] observed the effect of Rh3+ catalyst on the combustion of droplets at medium temperature and atmospheric pressure in experiments using crude vegetable oils (coconut oil-CCO, jatropha oil-CJO, and sunflower oil-CSFO). The oil droplets are multicomponent as they contain elements such as monounsaturated fatty acids, polyunsaturated fatty acids, saturated fatty acids, and glycerol. Experiments were conducted by suspending a droplet of vegetable oil with a diameter of about 0.6–1.1 mm at the junction of a thermocouple made of Pt/Rh13% and igniting it with a Ni-Cr electric coil heater. As a result of the experiments, it was found that the droplet diameter of all droplets increased as the heating time increased in catalyst-free combustion processes. However, when the catalyst was added, the diameter changes of crude coconut oil and crude sunflower seed droplets showed the same trend, while the crude jatropha oil droplet showed a different trend (**Figure 10**). **Table 3** shows the ignition times and diameters of the droplets at the moment of ignition.

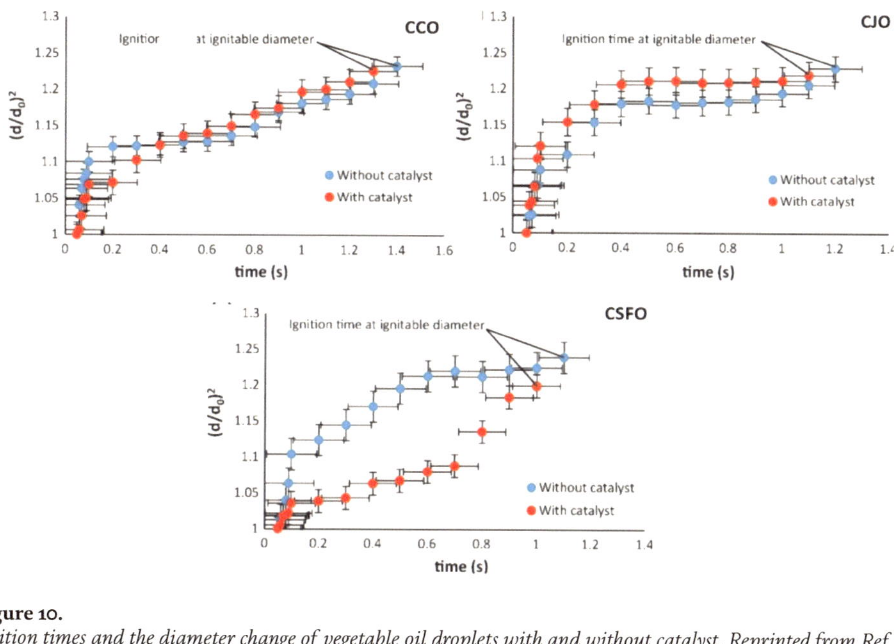

Figure 10.
Ignition times and the diameter change of vegetable oil droplets with and without catalyst. Reprinted from Ref. [38].

Ignition parameters	CCO		CJO		CSFO	
	With catalyst	Without catalyst	With catalyst	Without catalyst	With catalyst	Without catalyst
Ignition time (s)	1.3	1.4	1.1	1.2	1	1.1
Ignitable diameter (mm)	1.225	1.232	1.219	1.229	1.199	1.239

Source: Reprinted from Ref. [38].

Table 3.
Ignition diameters and ignition times of crude vegetable oil droplets.

Another study investigating the properties of jatropha oil droplets is by Wardana [39]. In the study, experiments were conducted by igniting a droplet of jatropha oil of different diameter and temperature at the junction of a thermocouple using an electric coil heater made of Ni-Cr wire. The experimental conditions are as follows:

Case 1: Droplet diameter 1.87 mm and temperature 30, 40, 50, 60, 70, 80°C.

Case 2: Droplet diameter 1.1, 1.22, 1.43, 1.56, 1.56, 1.75, 1.87 mm and temperature 70°C.

Combustion took place in two stages, with oil acid burning in the first stage and glycerol burning in the second stage. According to the obtained data, the oil temperature does not change the ignition temperature, but the diameter of the droplet does. A greater droplet diameter results in a lower ignition temperature.

Li et al. [40] experimentally investigated the combustion properties of ADN-based liquid propellant droplets in different atmospheres (oxidizing gas and inert gas – Ar, N_2O, NO, and O_2 atmospheres). In the study, the effect of parameters such as atmosphere, load voltage, ignition energy of droplet volume, ignition delay time, and combustion time on droplet equivalent diameter development during combustion was investigated. According to the results obtained in the experiments using the suspension method, an increase in droplet volume led to an increase in ignition delay time and combustion time. Atmosphere type had a significant effect on burning

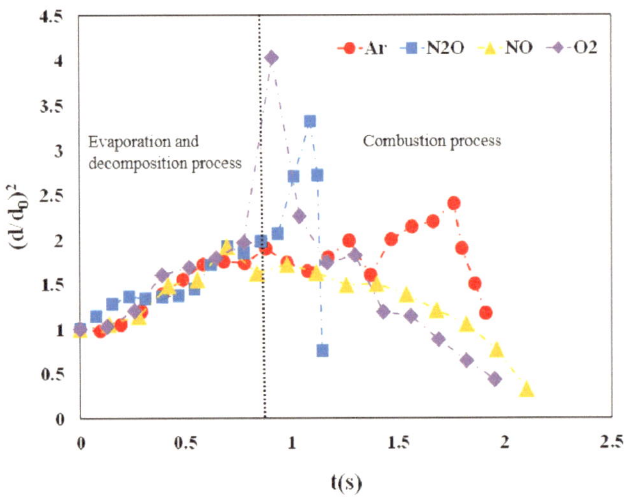

Figure 11.
The history of the squared diameter of ADN-based liquid propellant droplets in Ar, N_2O, NO, and O_2 atmospheres. Reprinted from Ref. [40].

time and droplet diameter but had little effect on ignition delay time. A comparison of the impact of various atmospheres on droplet diameter is presented in **Figure 11**. Accordingly, it appears that there is little difference in droplet diameter in different atmospheres during the evaporation and weathering stage. However, the combustion process was found to be more severe under N_2O and O_2 atmospheres, especially under N_2O atmosphere, but milder under NO atmosphere.

Yang et al. [41] studied the spray combustion properties of cedar sawdust bio-oil and kerosene. In the study, air for fuel droplet combustion was introduced into the combustion chamber from two axial sides of the combustor. Air inlet velocities were kept variable and fuel flow rate was kept constant. The turbulence thus created increases the effect of vaporizing droplets and mixing of the oxidizer, leading to accelerated combustion. As a result of the study, it was found that the droplet sizes in the center axis decreased when the air flow rate increased, but the droplet sizes outside the spray flow area did not change significantly with the airflow rate. Park et al. [42] conducted a study investigating the droplet atomization, fuel properties, combustion performance, and exhaust emission characteristics of gasoline-diesel direct blend fuels in a diesel engine. A phase Doppler particle analyzer was utilized to measure the droplet size of the fuels used in the study.

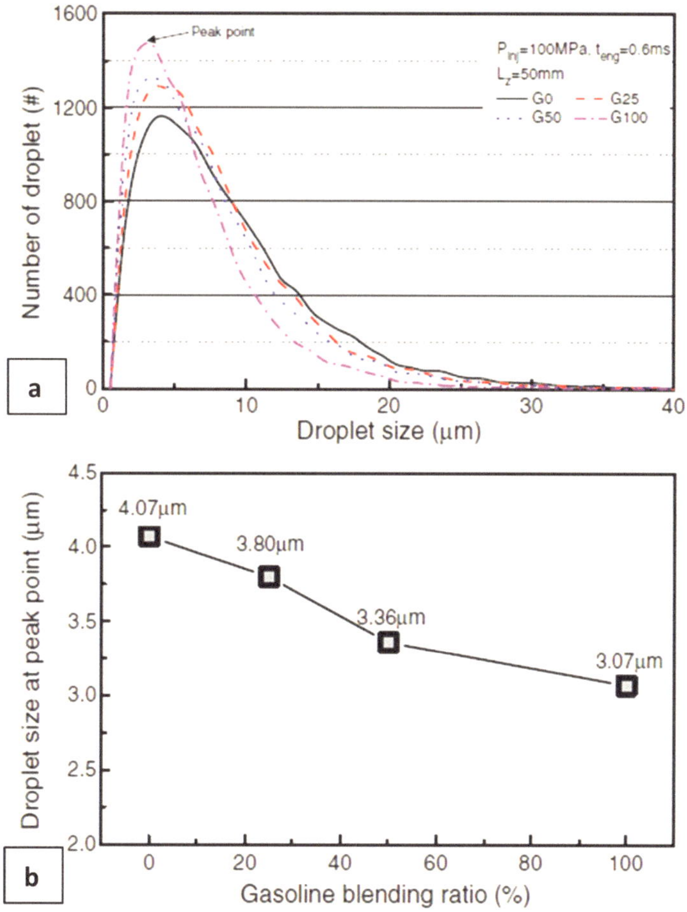

Figure 12.
Droplet number distribution (a) and droplet size at peak point for each test fuel (b). Reprinted from Ref. [42].

The combustion and emission properties of the fuels were examined using a diesel engine fitted with an emission analyzer and a common rail injection system. In an internal combustion engine, the increase of small droplets leads to an increase of the droplet surface in the same volume of injected spray, which leads to an increase in heat and energy transfer and is therefore an important parameter. **Figure 12** shows the droplet number distribution of gasoline-blended diesel fuels according to the average diameter. As can be seen from the figure, the number of small droplets increased as the gasoline concentration increased. Thus, the number of large droplets decreased, and the total average droplet size decreased. The droplet diameter changed to 3.80, 3.36, and 3.07 μm when the gasoline content rose to 25, 50, and 100%. As a result, it was discovered that pure diesel fuel had larger fuel droplets than gasoline-blended diesel fuel.

Farouk and Dryer [43], numerically investigated the two-stage combustion of n-heptane droplets in microgravity. The study used a transient, spherically symmetric droplet combustion model with detailed chemistry at both low and high temperature. It was discovered that the process and change of heat loss with the experimental setup are the primary parameters governing the two-stage combustion behavior shown in the tests. It was discovered that significant heat loss during first-stage combustion controls the transition to the second stage for medium-sized droplets in air. Diffusion/conductivity was shown to be the cause of significant heat loss in the second stage. Droplets with $d_0 > 3.0$ will result in two-stage combustion, according to parametric studies. For small droplet diameters, classical single-stage combustion was observed, while for sufficiently large droplets a dual combustion process was observed before extinction. The detailed model is qualitatively close to the experimental measurements, but the second-stage burning rate is overestimated. The second-stage extinction diameter was found to be ~0.35 mm, 3.5 times smaller than the experiment.

In the study of Xiao et al. [19], which was mentioned in the temperature section, examining the molecular dynamics simulations of a suspended n-dodecane fuel droplet under sub- and supercritical conditions, the evaporation properties of the droplet were also quantitatively determined depending on the droplet size. In the study,

Reference	Operating range	Fuel	Method
Ghamari and Ratner [37]	0.77–0.88 mm	Jet-A doped with carbon-based nanomaterials	Experimental/suspension
Nanlohy et al. [38]	1–1.3 mm	Crude vegetable oils with Rh^{3+} catalyst (CCO, CJO, CSFO)	Experimental/suspension
Wardana [39]	1.1–1.87 mm	Jatropha oil	Experimental/suspension
Li et al. [40]	1–4 mm	ADN-based liquid propellant	Experimental/suspension
Yang et al. [41]	$10–10^3$ μm	Cedar sawdust bio-oil and kerosene	Experimental/spray combustion
Park et al. [42]	0–40 μm	Gasoline-diesel direct blend	Experimental/injection
Farouk and Dryer [43]	2–7 mm	n-Heptane	Numerical/two-stage combustion
Xiao et al. [19]	21, 30, 43 nm	n-Dodecane	Numerical/suspension
Betelin et al. [44]	1, 5, 10, 50, 100, 500, 1000 μm	Hydrocarbon	Numerical/jet injection

Table 4.
Studies analyzed in the droplet size section.

the evaporation properties of the droplet are quantitatively determined depending on the ambient conditions and droplet size. Three droplets with initial diameters of approximately 21, 30, and 43 nm were simulated under the same conditions. Consequently, it was discovered that as ambient pressure and/or temperature rise, droplet lifespan falls. The evaporation rate constants for different droplets are quite close, but the larger droplet has a shorter heating time. Betelin et al. [44] conducted a simulation study on the non-equilibrium vaporization and ignition of small droplets of hydrocarbon fuel by liquid fuel jet injection into the hot combustion chamber atmosphere. In the study, it was observed that for small droplets (1, 5, 10 and 50 μm), the temperature initially increased due to external heating. Due to the rapid decrease in the relative velocity of the droplet, the evaporation in the gas flow decreases the temperature. Then, the decrease in droplet radius leads to an increase in heat fluxes and the temperature increases until the droplet disappears. For larger droplets (100, 500 and 1000 μm) the initial increase in temperature is higher, this relative velocity decreases much more slowly than for smaller droplets. As a result, a larger droplet is found to have a shorter lifetime (**Table 4**).

4. Conclusions

The study of the evaporation and combustion properties of fuel droplets allows a number of developments. These can be listed as more efficient use of fuels, increasing the efficiency of the combustion process, and reducing emissions. This review is a compilation of experimental and numerical studies on droplet evaporation and combustion characteristics of various fuels. The effects of three different parameters (temperature, pressure, droplet diameter) on the evaporation and combustion phases are investigated for different fuels, methods, and operating conditions. In this study, it is aimed to present a large number of various studies on this subject to the attention of researchers. The behavior of droplets of different fuel types under various combustion conditions and the results obtained were presented. Accordingly, the main results obtained are summarized below:

- Temperature in the evaporation and combustion environment is an important parameter affecting droplet lifetime. In general, the increase in temperature contributes to the shortening of droplet lifetime and acceleration of evaporation.

- The increase in ambient pressure leads to an increase in the lifetime of the fuel droplets and a decrease in the evaporation rate. The interaction of pressure with other parameters is as follows. While pressure increases the effect of turbulence, the effect of pressure decreases when the temperature reaches a certain point.

- If the droplet size is larger or smaller than it should be, vaporization delay or non-evaporation problems arise. Droplet size is affected by parameters such as temperature, air flow rate and has a significant effect on the combustion rate.

- In experimental studies on droplet evaporation, suspension method was mostly used. Methods can be developed and diversified. The fuel types and mixtures used can be diversified and enriched. Numerical studies on droplet evaporation are limited. Advanced modeling and simulation studies will be useful for experimental and application studies.

Author details

Saliha Özarslan[1] and Selçuk Sarıkoç[2]*

1 Automotive Technology Program, Vocational School of Technical Sciences, Amasya University, Amasya, Turkey

2 Department of Mechanical Engineering, Faculty of Engineering, Amasya University, Amasya, Turkey

*Address all correspondence to: sarikocselcuk@gmail.com

IntechOpen

© 2025 The Author(s). Licensee IntechOpen. This chapter is distributed under the terms of the Creative Commons Attribution License (http://creativecommons.org/licenses/by/4.0), which permits unrestricted use, distribution, and reproduction in any medium, provided the original work is properly cited.

References

[1] Som S, Datta A. Thermodynamic irreversibilities and exergy balance in combustion processes. Progress in Energy and Combustion Science. 2008;**34**(3):351-376. DOI: 10.1016/j.pecs.2007.09.001

[2] Ağbulut Ü, Sarıdemir S. A general view to converting fossil fuels to cleaner energy source by adding nanoparticles. International Journal of Ambient Energy. 2021;**42**(13):1569-1574. DOI: 10.1080/01430750.2018.1563822

[3] Perera F, Nadeau K. Climate change, fossil-fuel pollution, and children's health. New England Journal of Medicine. 2022;**386**(24):2303-2314. DOI: 10.1056/NEJMra2117706

[4] Variny M, Varga A, Rimár M, Janošovský J, Kizek J, Lukáč L, et al. Advances in biomass co-combustion with fossil fuels in the European context: A review. PRO. 2021;**9**(1):100. DOI: 10.3390/pr9010100

[5] Cherwoo L, Gupta I, Flora G, Verma R, Kapil M, Arya SK, et al. Biofuels an alternative to traditional fossil fuels: A comprehensive review. Sustainable Energy Technologies and Assessments. 2023;**60**:103503. DOI: 10.1016/j.seta.2023.103503

[6] Martins J, Brito F. Alternative fuels for internal combustion engines. Energies. 2020;**13**(16):4086. DOI: 10.3390/en13164086

[7] Holechek JL, Geli HM, Sawalhah MN, Valdez R. A global assessment: Can renewable energy replace fossil fuels by 2050? Sustainability. 2022;**14**(8):4792. DOI: 10.3390/su14084792

[8] Leach F, Kalghatgi G, Stone R, Miles P. The scope for improving the efficiency and environmental impact of internal combustion engines. Transportation Engineering. 2020;**1**:100005. DOI: 10.1016/j.treng.2020.100005

[9] Mehrizi AA, Karimi-maleh H, Naddafi M, Karaman O, Karimi F, Karaman C, et al. Evaporation characteristics of nanofuel droplets: A review. Fuel. 2022;**319**:123731. DOI: 10.1016/j.fuel.2022.123731

[10] Aggarwal SK. Single droplet ignition: Theoretical analyses and experimental findings. Progress in Energy and Combustion Science. 2014;**45**:79-107. DOI: 10.1016/j.pecs.2014.05.002

[11] Zhu S, Hu B, Akehurst S, Copeland C, Lewis A, Yuan H, et al. A review of water injection applied on the internal combustion engine. Energy Conversion and Management. 2019;**184**:139-158. DOI: 10.1016/j.enconman.2019.01.042

[12] Hoang AT. Combustion behavior, performance and emission characteristics of diesel engine fuelled with biodiesel containing cerium oxide nanoparticles: A review. Fuel Processing Technology. 2021;**218**:106840. DOI: 10.1016/j.fuproc.2021.106840

[13] Lee CC, Tran M-V, Tan BT, Scribano G, Chong CT. A comprehensive review on the effects of additives on fundamental combustion characteristics and pollutant formation of biodiesel and ethanol. Fuel. 2021;**288**:119749. DOI: 10.1016/j.fuel.2020.119749

[14] Téré D, Christian TG, Kayaba H, Boubou B, Sayouba S, Tizane D, et al. Evaporation and combustion of a drop of liquid fuel—A review. Smart Grid and Renewable Energy. 2022;**13**(2):28-54. DOI: 10.4236/sgre.2022.132003

[15] Sazhin SS. Modelling of fuel droplet heating and evaporation: Recent results and unsolved problems. Fuel. 2017;**196**:69-101. DOI: 10.1016/j.fuel.2017.01.048

[16] Wang Z, Yuan B, Huang Y, Cao J, Wang Y, Cheng X. Progress in experimental investigations on evaporation characteristics of a fuel droplet. Fuel Processing Technology. 2022;**231**:107243. DOI: 10.1016/j.fuproc.2022.107243

[17] Han K, Yang B, Zhao C, Fu G, Ma X, Song G. Experimental study on evaporation characteristics of ethanol–diesel blend fuel droplet. Experimental Thermal and Fluid Science. 2016;**70**:381-388. DOI: 10.1016/j.expthermflusci.2015.10.001

[18] Raghavan V. Numerical modeling of evaporation and combustion of isolated liquid fuel droplets: A review. Journal of the Indian Institute of Science. 2019;**99**(1):5-23

[19] Xiao G, Luo KH, Ma X, Shuai S. A molecular dynamics study of fuel droplet evaporation in sub-and supercritical conditions. Proceedings of the Combustion Institute. 2019;**37**(3):3219-3227. DOI: 10.1016/j.proci.2018.09.020

[20] Birouk M, Abou Al-Sood MM, Gökalp I. Droplet evaporation in a turbulent environment at elevated pressure and temperature conditions. Combustion Science and Technology. 2008;**180**(10-11):1987-2014. DOI: 10.1080/00102200802269707

[21] Ma X, Zhang F, Han K, Yang B, Song G. Evaporation characteristics of acetone–butanol–ethanol and diesel blends droplets at high ambient temperatures. Fuel. 2015;**160**:43-49. DOI: 10.1016/j.fuel.2015.07.079

[22] Javed I, Baek SW, Waheed K. Evaporation characteristics of heptane droplets with the addition of aluminum nanoparticles at elevated temperatures. Combustion and Flame. 2013;**160**(1):170-183. DOI: 10.1016/j.combustflame.2012.09.005

[23] Javed I, Baek SW, Waheed K. Autoignition and combustion characteristics of heptane droplets with the addition of aluminium nanoparticles at elevated temperatures. Combustion and Flame. 2015;**162**(1):191-206. DOI: 10.1016/j.combustflame.2014.07.015

[24] Javed I, Baek SW, Waheed K, Ali G, Cho SO. Evaporation characteristics of kerosene droplets with dilute concentrations of ligand-protected aluminum nanoparticles at elevated temperatures. Combustion and Flame. 2013;**160**(12):2955-2963. DOI: 10.1016/j.combustflame.2013.07.007

[25] Javed I, Baek SW, Waheed K. Autoignition and combustion characteristics of kerosene droplets with dilute concentrations of aluminum nanoparticles at elevated temperatures. Combustion and Flame. 2015;**162**(3):774-787. DOI: 10.1016/j.combustflame.2014.08.018

[26] Yu Y, Mei D, Gao Y, Qi J, Zhang C. Droplet evaporation characteristics of hydrogenated biodiesel-ethanol-diesel ternary fuel. Combustion Science and Technology. 2024;**196**(15):3576-3591. DOI: 10.1080/00102202.2023.2180633

[27] Hashimoto N, Nomura H, Suzuki M, Matsumoto T, Nishida H, Ozawa Y. Evaporation characteristics of a palm methyl ester droplet at high ambient temperatures. Fuel. 2015;**143**:202-210. DOI: 10.1016/j.fuel.2014.11.057

[28] Tao C, Miao K, Tang F, Qian Y, Zhang Y, Meng S, et al. Evaporation

and combustion characteristics for a droplet of lubricating oil and gasoline. International Communications in Heat and Mass Transfer. 2021;**127**:105513. DOI: 10.1016/j.icheatmasstransfer.2021.105513

[29] Verwey C, Birouk M. Fuel vaporization: Effect of droplet size and turbulence at elevated temperature and pressure. Combustion and Flame. 2018;**189**:33-45. DOI: 10.1016/j.combustflame.2017.10.010

[30] Birouk M, Fabbro SC. Droplet evaporation in a turbulent atmosphere at elevated pressure–experimental data. Proceedings of the Combustion Institute. 2013;**34**(1):1577-1584. DOI: 10.1016/j.proci.2012.05.092

[31] Birouk M. A hydrocarbon droplet evaporating in turbulent environment at elevated ambient pressure and temperature. Combustion Science and Technology. 2014;**186**(10-11):1295-1308. DOI: 10.1080/00102202.2014.934577

[32] Ma Z, Li Y, Li Z, Du W, Yin Z, Xu S. Evaporation and combustion characteristics of hydrocarbon fuel droplet in sub-and super-critical environments. Fuel. 2018;**220**:763-768. DOI: 10.1016/j.fuel.2018.02.073

[33] Birouk M, Toth SL. Hydrocarbon droplet turbulent combustion in an elevated pressure environment. Flow, Turbulence and Combustion. 2015;**94**:843-858. DOI: 10.1007/s10494-015-9601-9

[34] Kitano T, Nishio J, Kurose R, Komori S. Effects of ambient pressure, gas temperature and combustion reaction on droplet evaporation. Combustion and Flame. 2014;**161**(2):551-564. DOI: 10.1016/j.combustflame.2013.09.009

[35] Kim H, Sung N. The effect of ambient pressure on the evaporation of a single droplet and a spray. Combustion and Flame. 2003;**135**(3):261-270. DOI: 10.1016/S0010-2180(03)00165-2

[36] Aggarwal S, Mongia H. Multicomponent and high-pressure effects on droplet vaporization. Journal of Engineering for Gas Turbines and Power. 2002;**124**(2):248-255. DOI: 10.1115/1.1423640

[37] Ghamari M, Ratner A. Combustion characteristics of colloidal droplets of jet fuel and carbon based nanoparticles. Fuel. 2017;**188**:182-189. DOI: 10.1016/j.fuel.2016.10.040

[38] Nanlohy HY, Wardana I, Hamidi N, Yuliati L, Ueda T. The effect of Rh3+ catalyst on the combustion characteristics of crude vegetable oil droplets. Fuel. 2018;**220**:220-232. DOI: 10.1016/j.fuel.2018.02.001

[39] Wardana I. Combustion characteristics of jatropha oil droplet at various oil temperatures. Fuel. 2010;**89**(3):659-664. DOI: 10.1016/j.fuel.2009.07.002

[40] Li H-M, Li G-X, Li L. Comparative investigation on combustion characteristics of ADN-based liquid propellants in inert gas and oxidizing gas atmospheres with resistive ignition method. Fuel. 2023;**334**:126742. DOI: 10.1016/j.fuel.2022.126742

[41] Yang S, Wu M, Hsu T. Spray combustion characteristics of kerosene/bio-oil part I: Experimental study. Energy. 2017;**119**:26-36. DOI: 10.1016/j.energy.2016.12.062

[42] Park SH, Youn IM, Lim Y, Lee CS. Influence of the mixture of gasoline and diesel fuels on droplet atomization, combustion, and exhaust emission

characteristics in a compression ignition engine. Fuel Processing Technology. 2013;**106**:392-401. DOI: 10.1016/j.fuproc.2012.09.004

[43] Farouk TI, Dryer FL. Isolated n-heptane droplet combustion in microgravity:"cool flames"–two-stage combustion. Combustion and Flame. 2014;**161**(2):565-581. DOI: 10.1016/j.combustflame.2013.09.011

[44] Betelin V, Smirnov N, Nikitin V, Dushin V, Kushnirenko A, Nerchenko V. Evaporation and ignition of droplets in combustion chambers modeling and simulation. Acta Astronautica. 2012;**70**:23-35. DOI: 10.1016/j.actaastro.2011.06.021

Chapter 2

Automotive Radar Systems: Architecture, Signal Processing, and Future Perspectives

Tai Fei

Abstract

This chapter provides an overview of automotive radar technology, emphasizing its role in advanced driver-assistance systems (ADAS) through real-time measurements of range, angle, and velocity. We summarize the evolution of radar waveforms—such as frequency-modulated continuous wave (FMCW), phase-modulated continuous wave (PMCW), orthogonal frequency division multiplexing (OFDM), and phase-coded-FMCW (PC-FMCW)—and their signal-processing techniques for range-Doppler estimation and angle-of-arrival determination. Challenges in urban scenarios, such as the need for 4D sensing (range, velocity, azimuth, and elevation) and rising computational demands on radar sensors, are addressed. We also discuss the emergence of machine learning in radar perception and concerns over mutual interference in dense environments. The chapter concludes by exploring advances in radar hardware and algorithms, emphasizing real-time capability, interference mitigation, and progress toward higher autonomy.

Keywords: automotive radar, ADAS, MIMO, interference mitigation, real-time processing

1. Introduction

Radar has emerged as a cornerstone of automotive environmental perception, delivering vital spatial information—such as range, angle, and relative velocity—and enabling vehicles to form a comprehensive view of their surroundings [1, 2]. Serving as an integral component of numerous ADASs, including lane change warning (LCW) and emergency brake assist (EBA), radar sensors are indispensable for both safety enhancements and advanced driver-assistance functions.

Most modern automotive radar systems operate in the 77 GHz frequency band [3], although some—for instance, Forvia/Hella's third- and fourth-generation radars—still rely on the 24 GHz band. The 24 GHz ISM band provides up to 200 MHz of bandwidth in the European Union and 250 MHz in the United States [4], while the 76–77 GHz band is reserved for long-range applications, and 77–81 GHz is designated for short-range use. Compared to LiDAR and cameras, millimeter-wave radar provides robust performance in adverse weather conditions and offers a cost-effective solution for large-scale adoption in modern vehicles [5].

IntechOpen

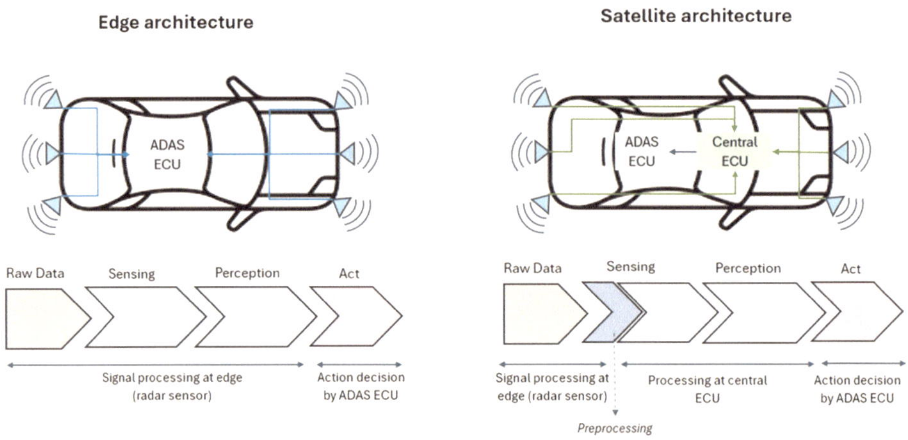

Figure 1.
Architecture used to integrate radar sensors into ADAS. Left: Edge architecture, signal processing including sensing and perception parts, is performed at the radar sensor. Right: Satellite architecture, radar sensor only performs a simple pre-processing and sends the data to the central ECU for data processing.

As the automotive industry advances toward higher autonomy levels (L-4 and L-5) [6], radar systems are subject to more stringent requirements for precision, reliability, and efficiency to address complex driving scenarios. Reflecting this evolution, radar sensors are increasingly integrated into ADAS architectures, illustrated in **Figure 1**. Under the conventional edge architecture, radar sensors process low-level data—encompassing object detection, classification, and tracking—and send compact high-level data *via* Controller Area Network (CAN) or a 100 Mb Ethernet interface to the ADAS electronic control unit (ECU). To unlock greater potential for sensor fusion and more accurate decision-making, the emerging satellite architecture streams pre-processed low-level data directly from the radar sensors to a powerful central ECU over a high-speed (often 1 GB Ethernet) link. This approach enhances scalability and modularity, facilitating flexible sensor placement and coverage around the vehicle. Consequently, a single platform can serve both cost-sensitive, low-end vehicles, and premium models, offering higher levels of autonomy.

In the following sections, we present an overview of radar technology, discuss its fundamental principles, and illustrate its roles and applications in ADASs.

2. Radar basics

2.1 Radar fundamentals

The foundational idea that metallic objects can reflect radio waves dates back to 1904, when the German engineer Christian Hülsmeyer patented (DE 165546) and demonstrated a device called the "Telemobiloskop." This early system was limited to detecting the presence of metallic objects (e.g., trains and ships) but could not measure their distance or speed. In June 1935, the Scottish radio engineer Sir Robert Watson-Watt—often cited as a key figure in the development of radar—successfully demonstrated the use of radio waves for detection and ranging. This achievement laid the groundwork for what would become known as "radar," an acronym for *Ra*dio *D*etection and *R*anging (**Figure 2**).

Figure 2.
Left: Christian Hülsmeyer. Right: Watson Watt.

Since those early breakthroughs, radar technology has seen considerable advances, particularly in the last two decades. Progress in semiconductor technology has enabled radar systems to operate at higher frequencies with improved sensitivity, while deep learning and other sophisticated signal-processing techniques have broadened radar's capabilities in target identification, tracking, and classification. As a result, modern radar systems now cover a broad spectrum of applications, including military surveillance, high-resolution imaging, traffic control, automotive driver assistance, and ground-penetrating radar for subsurface exploration. Beyond merely measuring distance, modern radar can extract target velocity, angle of arrival (AoA), radar cross section (RCS), and even classify types of movement or objects.

2.2 Principles of radar operation

In radar theory, a target is often modeled as a small point—sometimes extended to a small cluster of point-like reflectors—while the propagation medium is assumed to be free space. A radar system transmits electromagnetic (EM) waves through a transmit antenna, and these waves travel outward until they encounter an electrically conductive target surface. Upon reflection, a portion of the energy (the echo) returns to the radar, where it is collected by the receive antenna.

The fundamental parameter in radar detection is the round-trip delay τ. When a target is located at a distance R from the radar, the *delay* is given by

$$\tau = 2R/c, \tag{1}$$

where c is the speed of light in free space. Accurately estimating this delay is one of the primary objectives in radar signal processing. The method used to determine τ in practice depends on the radar's waveform design (i.e., its modulation scheme), which will be discussed in detail in Section 3.

Let P_s denote the transmitted power, and G_s the gain of the transmit antenna. At a distance R from the antenna, the directed power density is $S_g = \frac{P_s}{4\pi R^2} G_s$. If σ is the RCS of the target—representing how effectively the target reflects the EM wave—then the reflected power at the target is $P_r = S_g \sigma$. Assuming the radar's receive antenna has an effective aperture A_w and that the same distance R is traversed twice (outward and return paths), the received power is given by the *radar equation:*

$$P_e = \frac{P_s G_s \sigma A_w}{(4\pi)^2 R^4}. \quad (2)$$

This equation in (2) provides critical insight into how much power is returned from a target and is fundamental for designing radar systems (e.g., determining required transmit power, antenna gain, and receiver sensitivity). In practical scenarios, one must account for additional power losses due to atmospheric absorption and scattering, which can vary significantly with frequency, humidity, rain rate, and other environmental factors. Higher frequencies generally suffer more attenuation. Moreover, specific "resonance" bands exist where attenuation can spike relative to neighboring frequencies, making frequency selection a key design choice for radar engineers. More details on attenuation characteristics can be found in Ref. [7]. The RCS is highly dependent on the aspect angle at which the radar wave strikes a target. Complex objects such as automobiles can have surfaces oriented in many different ways, causing the σ values to fluctuate dramatically. According to measurements from the Remote Sensing Technology Institute at the German Aerospace Center (DLR) [8], RCS variations of up to 30 dBsm can be observed for vehicle targets. This variability underscores the importance of understanding how a target's geometry and orientation influence its detectability.

One of the key physical phenomena in radar technology is the Doppler effect. Initially formulated by Christian Doppler in 1842 for acoustic waves, it also applies to EM waves under standard conditions and is central to velocity measurements in radar systems. In essence, when there is relative motion between a radar sensor and a target, the transmitted EM wave undergoes a frequency shift. This shift, known as the Doppler shift, can be quantified using Chen's derivation in Ref. [9]:

$$f' = -\frac{v_{rel} f}{c}, \quad (3)$$

where f is the transmitted frequency, c is the speed of light, and v_{rel} is the relative velocity between the radar and the target. A positive value of v_{rel} indicates that the target is moving closer to the radar, resulting in an increase in perceived frequency.

3. Radar waveform and the associated signal processing

Rather than transmitting electromagnetic (EM) waves directly into free space, the radar signal is first *modulated* and then shifted to a specific carrier frequency, f_c. This modulation is crucial in practical radar scenarios because it provides timing marks or other signatures that enable the detection, localization, and characterization of targets. The choice of f_c and the corresponding bandwidth limitations are usually governed by local regulatory agencies.

One of the most straightforward waveforms for radar modulation is *pulse modulation*. In pulse-modulated radars, however, the requirement for high-range resolution often results in reduced radar sensitivity. To partially address this trade-off, the pulse can be extended using linear frequency modulation (LFM) and subsequently processed with a matched filter (see [10]). Although this strategy mitigates the conflict, it is best suited for high-power, long-range applications and requires high-dynamic-range receivers as well as powerful transmitters. Moreover, pulse modulation inherently involves *blind periods*, during which the receiver is switched off while the system transmits the next pulse.

In automotive radar applications, hardware simplicity is imperative for cost-effectiveness. As a result, state-of-the-art automotive radar sensors typically use low-cost microwave components, such as monolithic microwave integrated circuit (MMIC), making them affordable for large-scale production. Additionally, the transmit and receive antennas are separated and function independently, allowing for extended transmission durations. Under these conditions, the radar can maintain acceptable sensitivity levels even when operating under strict power regulations.

3.1 Signal model

In this subsection, we introduce a *waveform-agnostic* signal model for automotive radar. In this framework, the radar transmits a series of identical waveforms, which may take the form of FMCW chirps, OFDM, PMCW, or, more recently, PC-FMCW coded chirps. Conventionally, the time axis *within* each waveform is referred to as the *fast-time*, which is primarily responsible for range measurement, while the time axis *across* multiple waveforms is known as the *slow-time*, which is used for velocity measurement. Following the formulation in Ref. [3], the transmit signal at radio frequency (RF) is expressed as

$$x_{\mathrm{RF}}(t) = x_{\mathrm{BB}}(t - n_{\mathrm{s}}T_{\mathrm{s}}) \exp\left(\jmath 2\pi f_c (t - n_{\mathrm{s}}T_{\mathrm{s}})\right), \tag{4}$$

where $x_{\mathrm{BB}}(t)$ represents the equivalent complex baseband (ECB) waveform defined on the interval $0 < t < T_{\mathrm{s}}$. Here, T_{s} is the pulse repetition interval (PRI), f_c is the carrier frequency, and the ECB waveform repeats periodically over the slow-time $t_{\mathrm{s}} = n_{\mathrm{s}}T_{\mathrm{s}}$, for $n_{\mathrm{s}} \in \{0, 1, \ldots, N_{\mathrm{s}} - 1\}$. The quantity N_{s} denotes the total number of ECB waveforms within each coherent processing interval (CPI). Without loss of generality, we consider a reflected signal from a moving point target located at the time-varying range

$$d(t) = d_0 + v_{\mathrm{rel}} t_{\mathrm{f}} + v_{\mathrm{rel}} t_{\mathrm{s}}, \tag{5}$$

where $t_{\mathrm{f}} = t - n_{\mathrm{s}}T_{\mathrm{s}}$ denotes the *fast-time* $(0 \leq t_{\mathrm{f}} < T_{\mathrm{s}})$, and $n_{\mathrm{s}} = \lfloor t/T_{\mathrm{s}} \rfloor$. The corresponding round-trip delay is

$$\tau(t) = \tau_0 + \tau_{\mathrm{v}}(t_{\mathrm{f}}) + \tau_{\mathrm{v}}(t_{\mathrm{s}}), \tag{6}$$

where $\tau_0 = 2d_0/c$, $\tau_{\mathrm{v}}(t_{\mathrm{f}}) = 2v\, t_{\mathrm{f}}/c$, and $\tau_{\mathrm{v}}(t_{\mathrm{s}}) = 2v\, t_{\mathrm{s}}/c$. Hence, the received RF signal is

$$y_{\mathrm{RF}}(t) = A_r x_{\mathrm{RF}}(t - \tau(t)), \tag{7}$$

where A_r denotes the complex amplitude, capturing propagation losses, channel attenuation, and the target's RCS. After down-conversion, the corresponding baseband signal becomes

$$y_{\mathrm{BB}}(t_{\mathrm{f}}, t_{\mathrm{s}}) = A_r\, x_{\mathrm{BB}}\left(t_{\mathrm{f}} - (\tau_0 + \tau_{\mathrm{v}}(t_{\mathrm{s}}))\right) \exp\left(\jmath 2\pi f_{\mathrm{D}}\, (t_{\mathrm{f}} + t_{\mathrm{s}})\right), \tag{8}$$

where $f_{\mathrm{D}} = -2f_c v_{\mathrm{rel}}/c$ is the Doppler frequency. The factor of two in f_{D} reflects the round-trip effect, as the radar echo experiences the Doppler shift twice.

3.2 Waveforms and range-Doppler processing

The radar signal processing (RSP) primarily involves estimating a variety of target parameters, such as range, velocity (i.e., Doppler), AoA, target strength (i.e., RCS), and additional meta-information that assists with target classification and recognition in subsequent signal processing. In the mainstream RSP chain—adopted by most radar suppliers for efficiency—range and velocity are jointly estimated during the target detection stage. This significantly reduces the data volume so that only a limited number of detected point targets are then passed along for angle estimation (see Section 3.3), thereby reducing the computational burden.

By choosing different ECB in (4), the radar modulation is defined. In automotive applications, the RSP differs among various modulation schemes primarily in how range is obtained. In what follows, we introduce three typical modulation schemes (i.e., waveforms), placing emphasis on FMCW since it is the most widely deployed form of automotive radar on the road.

3.2.1 FMCW radar

The ECB waveform of a typical FMCW is described by the equation:

$$x_{\mathrm{BB}}(t_{\mathrm{f}}) = \exp\left(\jmath\pi\alpha t_{\mathrm{f}}^2\right), \tag{9}$$

where $\alpha = B/T_s$ represents the linear chirp slope, with B in the context of FMCW denoting the chirp *sweep bandwidth*. In automotive radar, B typically ranges from 100 MHz to 1 GHz at the 77 GHz band. Following Ref. [3], the reflected signal from a point target, after down-conversion, can be expressed as

$$y_{\mathrm{BB}}(t_{\mathrm{f}}, t_{\mathrm{s}}) = A_r \exp\left(\jmath\pi\alpha(t_{\mathrm{f}} - \tau_0 - \tau_{\mathrm{v}}(t_{\mathrm{s}}))^2\right) \exp\left(2\pi f_{\mathrm{D}}(t_{\mathrm{f}} + t_{\mathrm{s}})\right). \tag{10}$$

This signal is then conjugately mixed with the reference waveform signal from (9) and passed through a lowpass filter (LPF), resulting in a *beat signal*,

$$y_{\mathrm{b}}(t_{\mathrm{f}}, t_{\mathrm{s}}) = A_r \exp\left(\jmath 2\pi f_{\mathrm{b}} t_{\mathrm{f}}\right) \exp\left(2\pi f_{\mathrm{D}}(t_{\mathrm{f}} + t_{\mathrm{s}})\right), \tag{11}$$

where $f_{\mathrm{b}} = \alpha\tau(t)$. It is worth noting that the term $\exp(-2\pi f_{\mathrm{D}} t_{\mathrm{f}})$ describes the Doppler shift of the waveform in fast-time. It has an adverse effect on the range estimation (also leads to intercarrier interference in OFDM) when not taken into account.

This beat signal, whose bandwidth is generally much narrower (mostly lower than 20 MHz) than the one of the baseband signal in (10), can be easily digitized by an analog-to-digital converter (ADC) and organized into a 2D matrix $\mathbf{Y}_{\mathrm{b}} = \left(y_{\mathrm{b}}[n_{\mathrm{f}}, n_{\mathrm{s}}]\right) \in \mathbb{C}^{N_{\mathrm{f}} \times N_{\mathrm{s}}}$, where $y_{\mathrm{b}}[n_{\mathrm{f}}, n_{\mathrm{s}}] = y_{\mathrm{b}}(n_{\mathrm{f}} T_{\mathrm{c}} + n_{\mathrm{s}} T_{\mathrm{s}})$. According to Ref. [11], the sampled beat signal $y_{\mathrm{b}}[n_{\mathrm{f}}, n_{\mathrm{s}}]$ can be formulated as

$$y_{\mathrm{b}}[n_{\mathrm{f}}, n_{\mathrm{s}}] = A_r \exp\left(\jmath 2\pi (f_{\mathrm{b}} + f_{\mathrm{D}}) n_{\mathrm{f}} T_{\mathrm{c}}\right) \exp\left(2\pi f_{\mathrm{D}} n_{\mathrm{s}} T_{\mathrm{s}}\right), \tag{12}$$

where n_{f} and n_{s} represent indices along the fast-time (t_{f}) and slow-time (t_{s}) dimensions, respectively. T_{c} is the sampling period, and $1/T_{\mathrm{c}}$ is the sampling rate. A common practice is to apply a discrete Fourier transform (DFT) along the fast-time dimension to obtain a *range profile* matrix, $\mathbf{R} = [\boldsymbol{rr}_0, \ldots, \boldsymbol{rr}_{N_s-1}] \in \mathbb{C}^{N_f \times N_s}$, in which the ranges of targets are present, then apply a second DFT along the slow-time dimension for velocity estimation. In short, a 2D DFT can be applied to $\mathbf{Y}_b$ to obtain the *range-Doppler (RD)* spectrum, $\mathcal{S}$:

$$\mathcal{S}[k_r, k_v] = \sum_{n_f=0}^{N_f-1} \sum_{n_s=0}^{N_s-1} y_b[n_f, n_s] \exp\left(-j2\pi\left(\frac{k_r}{N_f}n_f + \frac{k_v}{N_s}n_s\right)\right), \tag{13}$$

where k_r and k_v denote the range and velocity bin indices, respectively. From this $\mathcal{S}$, range and velocity are obtained by mapping these indices to physical units. The respective resolutions, Δr and Δv_{rel}, follow [11] as

$$\begin{cases} \Delta r = c \cdot \dfrac{T_c}{2B} \cdot \Delta f_r = \dfrac{c}{2B} \\ \Delta v_{rel} = \dfrac{\lambda_c}{2} \cdot \Delta f_D \end{cases}, \tag{14}$$

where $\lambda_c = c/f_c$, , $\Delta f_r = 1/T_c$, and $\Delta f_D = 1/(N_s T_s)$. The RD processing for FMCW radar is summarized in **Figure 3**. In many automotive radar systems, constant false alarm rate (CFAR) thresholding [12] (or one of its variants) is applied to the magnitude of the RD spectrum to detect targets, typically by identifying local maxima (i.e., *peaks*) exceeding a specified threshold. The range and relative velocity are then obtained by multiplying the peak indices in $\mathcal{S}$ by the corresponding resolution values in (14). Because the velocity index can take negative values, an index offset $k_{v,0}$ (which represents $f_D = 0$) must be subtracted from k_v before this multiplication.

In practice, a window function (e.g., Hamming or Blackman) is typically applied in (13) to mitigate the spectral leakage inherent in the DFT. Furthermore, to enhance the

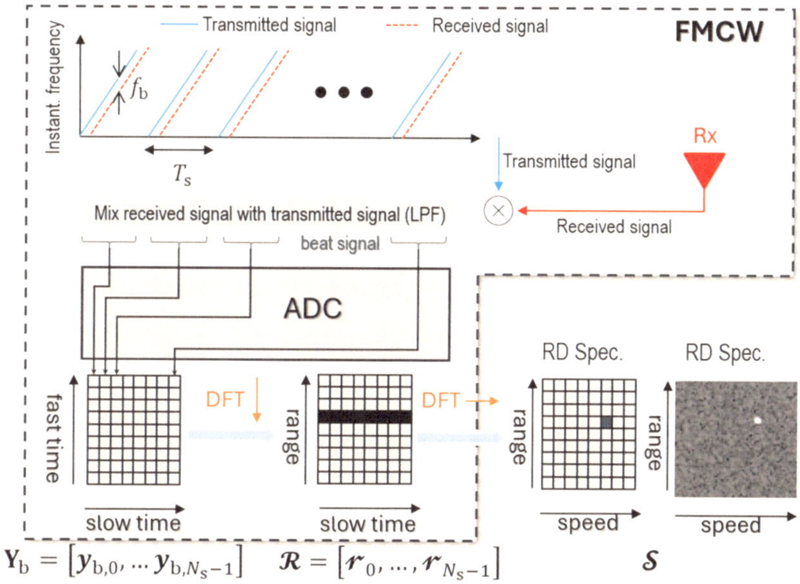

Figure 3.
The range-Doppler processing chain of FMCW modulation. The part inside the dashed line block is FMCW-specific processing.

accuracy of range and velocity estimates—otherwise limited by resolution—*parabolic interpolation* can be employed to locate non-integer peak indices that lie between two adjacent frequency bins. More recently, various high-resolution techniques have been proposed, such as the matrix-pencil-based target separation in [11], to address scenarios where multiple point targets partially overlap in the RD spectrum.

3.2.2 PMCW radar

The baseband signal of a PMCW radar is a pure phase-coded waveform segmented into N_{c_1} time intervals, commonly referred to as chips, each assigned a distinct phase. To simplify hardware implementation, binary phase shift keying (BPSK) modulation is typically employed, wherein the phase sequence elements satisfy $\varphi_{n_c} \in \{0, \pi\}$ for each chip n_c, with N_{c_1} representing the sequence length. The ECB representation of PMCW is expressed as

$$\begin{aligned} x_{\mathrm{BB}}(t_{\mathrm{f}}) &= s_1(t_{\mathrm{f}}) \\ &= \sum_{n_c=0}^{N_{c_1}-1} \exp(\jmath\varphi_{n_c})\,\mathrm{rect}\left(\frac{t_{\mathrm{f}} - n_c T_{c_1}}{T_{c_1}}\right), \end{aligned} \tag{15}$$

where $\mathrm{rect}\left(\frac{t}{T_c}\right)$ denotes a rectangular window function, T_{c_1} is the chip duration, and $0 \leq n_c < N_{c_1}$. Similar to FMCW, N_s consecutive identical sequences are transmitted within a single CPI. In the context of PMCW, the *instantaneous bandwidth* of the transmitted sequence is determined by the chip duration, given by $B = 1/T_{c_1}$, which directly influences the range resolution [13].

At the receiver, the baseband signal $y_{\mathrm{BB}}(t_{\mathrm{f}}, t_{\mathrm{s}})$ is sampled by an ADC, and according to Ref. [14], the sampling frequency must be at least $1/T_{c_1}$. Unlike the sampling of the beat signal in Eq. (11), sampling $y_{\mathrm{BB}}(t_{\mathrm{f}}, t_{\mathrm{s}})$ imposes more stringent requirements on the ADC because its instantaneous bandwidth is typically much wider than that of $y_{\mathrm{b}}(t_{\mathrm{f}}, t_{\mathrm{s}})$. This broader bandwidth is necessary to achieve the desired range resolution, which results in a very short T_{c_1}. Recently, Kahlert et al. [14] proposed a variant of PMCW that employs stepped carrier frequencies across the slow-time dimension. This approach effectively reduces the required sampling rate along the fast-time axis while maintaining comparable range resolution. By varying the carrier frequency in each slow-time interval, the instantaneous bandwidth can be managed more efficiently, alleviating the stringent ADC requirements without compromising performance.

The sampled receive signal is stored in a 2D matrix $\mathbf{Y}_{\mathrm{BB}} = (y_{\mathrm{BB}}[n_{\mathrm{f}}, n_{\mathrm{s}}]) \in \mathbb{C}^{N_{c_1} \times N_s}$. To extract range information, the correlation between the transmitted sequence and the received signals is computed, producing a range profile. Subsequently, a DFT is applied to the range profile along the slow-time dimension to determine the Doppler frequency, corresponding to the relative velocity v_{rel}.

3.2.3 OFDM radar

The OFDM scheme, commonly recognized as a multi-carrier modulation technique, utilizes coding in the frequency domain. An OFDM radar signal in the baseband consists of N_s symbols. Within the frequency domain, the bandwidth of each OFDM symbol is divided into N_b subcarriers, each with a bandwidth of Δf. Each subcarrier carries a code symbol $c_{m_\mathrm{f},n_\mathrm{s}}$, drawn from a predefined modulation alphabet such as quadrature phase shift keying (QPSK) or quadrate amplitude modulation (QAM). The baseband signal is then expressed as follows:

$$x_\mathrm{BB}(t_\mathrm{f}) = \frac{1}{N_\mathrm{b}} \sum_{m_\mathrm{f}=0}^{N_\mathrm{b}-1} c_{m_\mathrm{f},n_\mathrm{s}} \exp(\jmath 2\pi (m_\mathrm{f}\Delta f) t_\mathrm{f}), \tag{16}$$

where $\Delta f = B/N_\mathrm{b}$ and B is the OFDM symbol bandwidth. In practical implementations, a cyclic prefix (CP) is inserted between consecutive OFDM symbols to prevent intersymbol interference. The received signal is first sampled by an ADC with a sampling frequency B. Subsequently, the CP is removed before storing the signal in the 2D matrix $\mathbf{Y}_\mathrm{BB} = (y_\mathrm{BB}[n_\mathrm{f}, n_\mathrm{s}]) \in \mathbb{C}^{N_\mathrm{b}\times N_\mathrm{s}}$. To eliminate the coding in the frequency domain, a DFT is applied to $\mathbf{Y}_\mathrm{BB}$ along the fast-time dimension, resulting in $\mathcal{Y}_\mathrm{BB} = (\tilde{y}_\mathrm{BB}[m_\mathrm{f}, n_\mathrm{s}]) \in \mathbb{C}^{N_\mathrm{b}\times N_\mathrm{s}}$:

$$\tilde{y}_\mathrm{BB}[m_\mathrm{f}, n_\mathrm{s}] = \tilde{A}_r c_{m_\mathrm{f},n_\mathrm{s}} \exp(-\jmath 2\pi\tau\Delta f m_\mathrm{f}) \exp(2\pi T_\mathrm{s} f_\mathrm{D} n_\mathrm{s}), \tag{17}$$

where $\tilde{A}_r$ denotes the complex amplitude scaling factor in the OFDM processing. Since the receiver has prior knowledge of the code symbols $c_{m_\mathrm{f},n_\mathrm{s}}$, these can be readily removed from Eq. (17) through element-wise division. RD processing is then performed by applying an inverse discrete Fourier transform (IDFT) along the m_f axis and a DFT along the n_s axis [15].

A summary of the range processing techniques for PMCW and OFDM is illustrated in **Figure 4**. Both modulation schemes leverage coding techniques derived from communication technologies and are therefore classified as *digital modulation* schemes. They have been extensively researched over the past decades as promising alternatives to FMCW. However, their widespread adoption in production remains limited due to the current maturity level of semiconductor technologies and the stringent cost-efficiency requirements of the automotive market.

3.2.4 PC-FMCW radar

PC-FMCW augments a chirp waveform with a phase-coded sequence along the fast-time axis. Its ECB representation can be written as

$$x_\mathrm{BB}(t_\mathrm{f}) = s_2(t_\mathrm{f}) \exp(\jmath \pi \alpha t_\mathrm{f}^2), \tag{18}$$

where $s_2(t_\mathrm{f})$ is a code sequence of N_{c_2} chips,

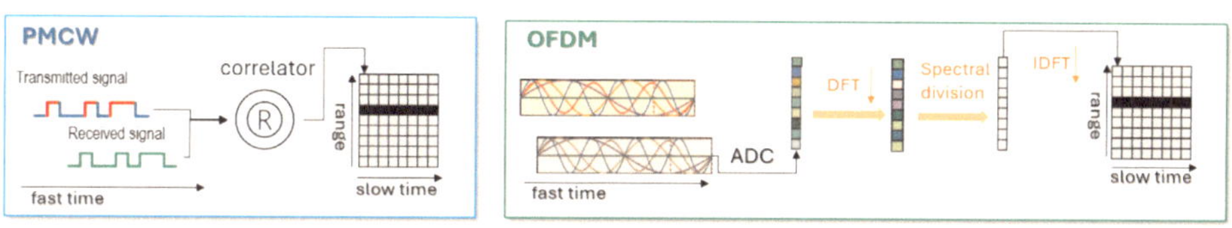

Figure 4.
Range processing. Left: range processing associated with PMCW. Right: range processing associated with OFDM.

$$s_2(t_f) = \sum_{n_c=0}^{N_{c_2}-1} \exp(\jmath\varphi_{n_c}) \operatorname{rect}\left(\frac{t_f - n_c T_{c_2}}{T_{c_2}}\right), \tag{19}$$

and T_{c_2} denotes the chip duration. Typically, $T_{c_2} \gg T_{c_1}$, meaning the chips in PC-FMCW are significantly longer than those in PMCW. As a result, PC-FMCW offers a *trade-off solution* for industry applications, relaxing ADC requirements while still enabling code-domain multiplexing.

For demodulation, the baseband signal from (19) is first multiplied by the complex conjugate of the uncoded chirp, $(\jmath\pi\alpha t_f^2)$, and then passed through a LPF. The resulting beat signal, which still contains the code sequence, is expressed as

$$y_{b,c}(t_f, t_s) = A_r s_2(t_f - \tau_0 - \tau_v(t_s)) \exp(\jmath 2\pi f_b t_f) \exp\left(\jmath 2\pi f_D(t_f + t_s)\right). \tag{20}$$

Prior to RD processing, the delayed code sequence must be removed, a process referred to as decoding. In Ref. [16], $s_2(t_f)$ is treated as a slowly varying envelope, and a group-delay filter is used to align each beat frequency's envelope. After decoding, range and velocity information can be extracted, as detailed in Ref. [16].

3.3 Angle estimation

A point target's range, relative velocity, and reflectivity can be extracted through RD processing and target detection. For most ADAS functions, however, additional spatial information is required—at least in two dimensions. Hence, determining the AoA, specifically the azimuth, is crucial. Drawing an analogy to human vision, using only one receive antenna (akin to a single eye) makes spatial perception difficult, so at least two receive antennas are needed to estimate AoA.

As depicted in **Figure 5**, the point target is assumed to lie in the far field, that is, its distance to the receiver array greatly exceeds λ_c. Consequently, the wavefront of its echo can be approximated as planar. The wavefront arrives first at the reference antenna Rx_0, then at the lth receive antenna Rx_l after a delay, where $0 \le l < L - 1$ and L denotes the number of receive antennas. In other words, the range measured at Rx_l is offset by $r_{bias,l} = dl \sin(\theta)$ relative to Rx_0, where d is the spacing between antennas and θ is the target's azimuth. Although this offset directly indicates azimuth, it is typically much smaller than the radar's range resolution, making it difficult to measure directly in the RD spectra (where all antennas see the same target position). Instead, one can exploit the phase difference among signals at different antennas, which roots in the range offset. Considering the signals at Rx_0 and Rx_l ($S_0[k_r, k_v]$ and $S_l[k_r, k_v]$), their phase difference ψ_l is

$$\psi_l = \frac{r_{bias,l}}{\lambda_c} \cdot 2\pi. \tag{21}$$

For simplicity, consider just two receive antennas, Rx_0 and Rx_1. Substituting $r_{bias,1} = d \sin(\theta)$ into (21) and rearranging yields

$$\theta = \arcsin\left(\frac{\psi_1 \lambda_c}{2\pi d}\right), \tag{22}$$

which is known as the *monopulse* AoA estimation method, predominantly used in scenarios with a single target. Note that the spacing d cannot be chosen arbitrarily. To ensure unambiguous $\theta \in (-\pi/2, \pi/2]$ for $\psi_1 \in (-\pi, \pi]$, we derive the upper bound

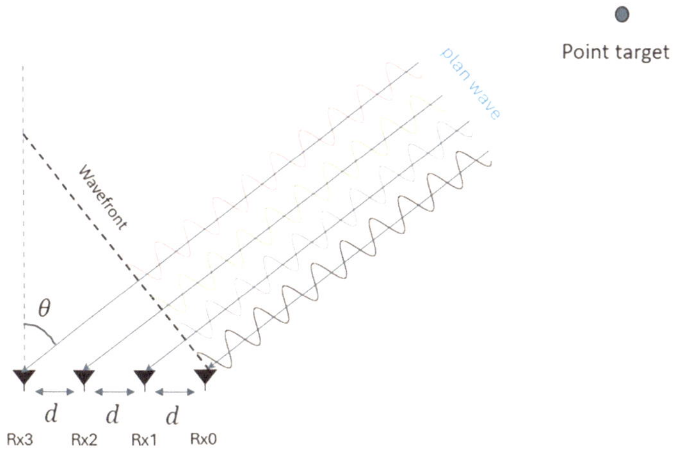

Figure 5.
Illustration of AoA estimation. In this example, four receive antennas are used, denoted by $Rx_0 - Rx_3$. *A point target is far enough away that its distance to each antenna is much larger than the carrier wavelength* λ_c.

$$d_{\max} = \frac{\lambda_c}{c}. \tag{23}$$

When multiple point targets are present, a single-phase difference is insufficient for resolving their distinct azimuths. Additional antennas, such as Rx_2 and Rx_3 in **Figure 5**, provide more phase measurements of the form $\psi_l = l \times \frac{2\pi d \sin(\theta)}{\lambda_c}$. Each subsequent ψ_l introduces an incremental phase shift relative to the previous antenna, forming a linear progression in phase analogous to a *frequency*. Commonly referred to as the angular frequency $\omega = \frac{2\pi d \sin(\theta)}{\lambda_c}$, this frequency can be found by performing a DFT across the antenna dimension for signals at the same RD coordinates. The bound $d_{\max}$ on antenna spacing is analogous to the Nyquist criterion: At least two spatial samples (receive antennas) must be contained within one *wavelength* λ_c to avoid spatial aliasing.

Equidistant receive antennas with $d < d_{\max}$ form a uniform linear antenna array (ULA). By increasing the number of receive antennas, the aperture of the ULA (i.e., its overall length) increases, which in turn improves the angular resolution—analogous to how extending the observation duration in a DFT refines frequency resolution. For example, with $d = \lambda_c/2$, a coarse estimation of the angular resolution θ_{res} is given by the rule of thumb

$$\theta_{\text{res}} = \frac{2}{L}. \tag{24}$$

Due to the strict size constraints associated with integrating automotive radar into a car body, it is infeasible to arbitrarily enlarge the physical antenna aperture to achieve the desired angular resolution. As a remedy, Multiple-Input, Multiple-Output (MIMO) antenna layouts are employed, leveraging multiple transmit (Tx) and receive antennas. In such systems, a single Rx antenna must distinguish echoes from all Tx antennas, often achieved *via* orthogonal waveforms or time-multiplexing. As illustrated in **Figure 6**, phase differences ψ_4 through ψ_7 in the physical aperture are practically created at the transmit side *via* antenna spacing between the Tx antennas.

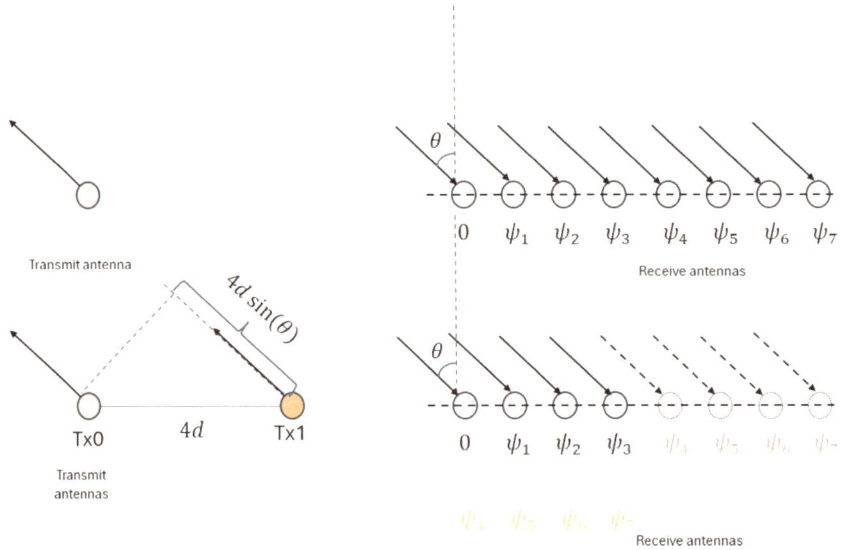

Figure 6.
MIMO antenna concept. Upper: a ULA with one transmit antenna and its physical aperture. Lower: a MIMO array with two Tx antennas, where the virtual array has the same aperture as the physical array shown in the upper figure.

A critical aspect in forming the virtual aperture in MIMO is ensuring that echoes from different Tx antennas remain separable at each Rx antenna. Orthogonal coding techniques, such as PMCW and OFDM, are frequently adopted for this purpose. Additionally, time-multiplexing, which enables Tx antennas to transmit their signals sequentially, is also widely employed. However, in highly dynamic scenarios, compensating for the phase shifts introduced by time-multiplexing poses significant challenges. Recently, Doppler division multiplexing [17]—where phases of the transmitted signal are encoded across slow-time—has been gaining traction among many MMIC suppliers as a standard solution for automotive MIMO.

3.4 Summary

In this section, we outline the primary modulation schemes employed in automotive radar systems. Subsequent radar signal processing (RSP) extracts critical target parameters, including range, velocity, and AoA, as well as additional metadata. As illustrated in **Figure** 7, these parameters are efficiently consolidated into a target list, which is then passed to post-processing modules [18], such as stationary and moving target processing.

Stationary target processing is typically applied in scenarios like parking to support occupancy grid mapping and obstacle classification. In contrast, moving target processing focuses on tasks such as object tracking and road-user classification. The outputs from both processing methods are integrated into the ADAS ECU, enabling various ADAS functions to determine appropriate actions.

Notably, when the radar sensor uses a single transmit antenna, the ADC data from a single receive antenna, denoted as $\mathbf{Y}_b$, is already channel-separated and can be directly forwarded to RD processing. However, in time-multiplexing-based MIMO systems, as depicted in **Figure** 7, the ADC data from a single receive antenna is a composite signal comprising contributions from multiple transmit antennas.

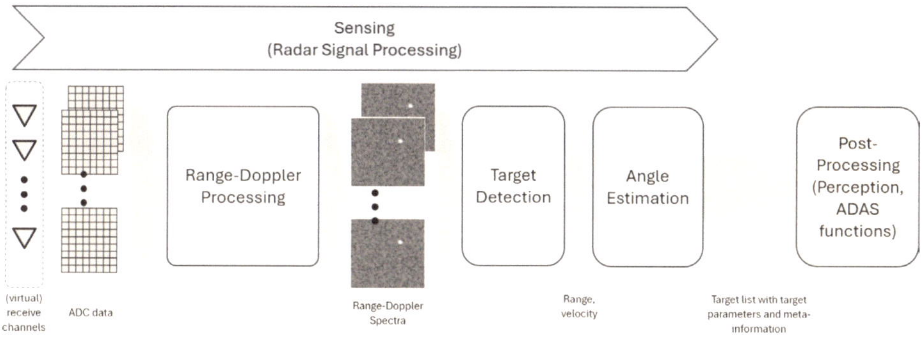

Figure 7.
RSP processing chain: In this configuration, the separation of ADC data into different virtual channels occurs prior to RD processing. After target detection, the data volume is significantly reduced, resulting in a concise target list.

This mixed data must be separated into distinct virtual channels. For instance, in a configuration with two transmit antennas (Tx) and four receive antennas (Rx), as shown in **Figure 6**, the separation process produces eight virtual channels.

4. Challenges and trends in real applications

The evolution of automotive radar is steadily advancing toward highly automated driving, imposing ever-greater demands on radar performance. These requirements are being addressed through both hardware innovations and algorithmic advancements. Moreover, *real-time capability* of RSP has become a critical challenge, especially as the complexity of radar signal-processing algorithms grows. This increasing algorithmic complexity significantly raises the computational load on embedded processors in radar sensors, which must operate under strict constraints on size, power consumption, and cost while still meeting stringent latency requirements.

To contend with increasingly complex urban scenarios, merely 2D spatial information (i.e., azimuth AoA only) is no longer sufficient. In recent years, many radar suppliers have been developing the so-called *4D radar* sensors, capable of estimating range, velocity, azimuth, and elevation for each point target. One common approach is to cascade multiple MMIC devices to maximize the virtual antenna array in a MIMO configuration. To further exploit the spatial dimension, sparse 2D MIMO arrays along azimuth and elevation are also being investigated, leveraging the inherent sparsity in the scene [19].

The recent success of machine learning and deep learning has likewise found utility in RSP [20] and in post-processing modules for environmental perception [18]. However, the scarcity of high-quality annotated radar data remains a formidable hurdle. Small changes in modulation parameters can significantly alter the radar returns, making data transferability weak. Moreover, expert knowledge is often required to interpret radar data, complicating the creation of extensive training sets. Ongoing work in the research community includes the development of realistic radar data simulations to address these data limitations [21], as well as cross-modal association strategies [22] to facilitate model training.

In parallel, there is heightened concern about mutual interference, particularly as autonomous vehicles multiply on public roads [23]. Large-scale governmental initiatives, such as the MOSARIM project [24] funded by the European Union

(2009–2012), identified the importance of integrating interference-mitigation measures into radar design. Subsequent research by the U.S. National Highway Traffic Safety Administration (NHTSA) and Germany's IMIKO-Radar project [25] continued to investigate automotive radar interference. Traditional mitigation approaches—through time, frequency, code, or spatial separation—have proven foundational. The IMIKO-Radar project, in particular, explored cooperative solutions to maintain radar accuracy under interference. Recently, both academic and industrial work [26, 27] has highlighted the growing importance of interference-resilient techniques for safe and reliable autonomous driving.

5. Conclusion and outlook

In summary, automotive radar technology has evolved into a crucial pillar of advanced driver-assistance systems, delivering key insights such as range, angle, and velocity under diverse driving and weather conditions. Today's radars rely on high-frequency operation in the millimeter-wave band, providing a robust, cost-effective sensor solution for modern vehicles. Both hardware and algorithmic innovations—especially in the realms of high-speed signal processing, MIMO antenna designs, and digital modulation schemes—continue to extend radar capabilities, paving the way toward higher levels of driving automation.

Looking ahead, several challenges and trends will shape the field. First, ongoing miniaturization and integration demand carefully balanced trade-offs between size, cost, and performance, which may spur further development of new semiconductor processes and 2D/3D MIMO antenna layouts. Second, advanced radar waveforms and real-time signal processing—boosted by machine learning—are poised to reduce latency and enhance environmental perception, although a critical bottleneck remains in the scarcity of high-quality annotated data. Addressing transferability issues and automating radar data annotation will be key to unlocking radar's full potential in learning-based approaches. Third, the proliferation of radar-equipped vehicles brings renewed attention to interference-mitigation strategies, compelling researchers and industries to collaborate on robust, cooperative approaches. The success of such efforts will determine the reliability and safety of autonomous driving in increasingly dense automotive environments.

Overall, continued synergy among semiconductor advancements, waveform design, and algorithmic innovation—coupled with the reliable handling of mutual interference—will be vital in overcoming these challenges. By building upon strong foundations in radar hardware and signal processing, the automotive industry is poised to deliver next-generation radar solutions that meet the demanding requirements of highly automated driving.

Abbreviations

ADC	analog-to-digital converter
ADAS	advanced driver-assistance system
LCW	lane change warning
EBA	emergency brake assist
CAN	controller area network
ECU	electronic control unit

AoA	angle of arrival
RCS	radar cross section
EM	electromagnetic
LFM	linear frequency modulation
MMIC	monolithic microwave integrated circuit
FMCW	frequency-modulated continuous wave
PMCW	phase-modulated continuous wave
OFDM	orthogonal frequency division multiplexing
PC-FMCW	phase-coded frequency-modulated continuous wave
ECB	equivalent complex baseband
PRI	pulse repetition interval
CPI	coherent processing interval
RF	radio frequency
RSP	radar signal processing
LPF	lowpass filter
RD	range-Doppler
DFT	discrete Fourier transform
CFAR	constant false alarm rate
BPSK	binary phase shift keying
QPSK	quadrature phase shift keying
QAM	quadrate amplitude modulation
IDFT	inverse discrete Fourier transform
CP	cyclic prefix
ULA	uniform linear array
MIMO	multiple-input, multiple-output

Author details

Tai Fei
Dortmund University of Applied Sciences and Arts, Dortmund, Germany

*Address all correspondence to: tai.fei@ieee.org

IntechOpen

© 2025 The Author(s). Licensee IntechOpen. This chapter is distributed under the terms of the Creative Commons Attribution License (http://creativecommons.org/licenses/by/4.0), which permits unrestricted use, distribution, and reproduction in any medium, provided the original work is properly cited.

References

[1] Dickmann J, Klappstein J, Hahn M, Appenrodt N, Bloecher H-L, Werber K, et al. Automotive radar the key technology for autonomous driving: From detection and ranging to environmental understanding. In: 2016 IEEE Radar Conference (RadarConf). Philadelphia, PA, USA: IEEE; 2016. pp. 1-6

[2] Kahlert M, Peitzmeier H, Evans D, Talits K, Kortmann F, Tebruegge C. Resilience of spatial environment perception towards fully automated driving: A review. IEEE Sensors Journal. 2024;**24**(14):1-1

[3] Hakobyan G, Yang B. High-performance automotive radar: A review of signal processing algorithms and modulation schemes. IEEE Signal Processing Magazine. 2019;**36**(5):32-44

[4] Won Y-S, Kim C-H, Lee S-G. Range resolution improvement of a 24 Ghz ism band pulse radar—A feasibility study. IEEE Sensors Journal. 2015;**15**(12): 7142-7149

[5] Fei T, Mukhopadhyay SC, Da Costa JPJ, Lan CR, Demitri N. Spatial environment perception and sensing in automated systems: A review. IEEE Sensors Journal. 2024;**24**(14): 21813-21833

[6] Khan MA et al. Level-5 autonomous driving—Are we there yet? A review of research literature. ACM Computing Surveys. 2022;**55**(2):1-38

[7] Tataria H, Shafi M, Molisch AF, Dohler M, Sjöland H, Tufvesson F. 6g wireless systems: Vision, requirements, challenges, insights, and opportunities. Proceedings of the IEEE. 2021;**109**(7): 1166-1199

[8] Palubinskas G, Runge H. Radar signatures of a passenger car. IEEE Geoscience and Remote Sensing Letters. 2007;**4**(4):644-648

[9] Chen VC. The Micro-Doppler Effect in Radar. Artech House Radar Library. Norwood, MA, USA: Artech House; 2011

[10] Letao X, Feng D, Wang X. Matched-filter properties of linear-frequency-modulation radar signal reflected from a phase-switched screen. IET Radar, Sonar & Navigation. 2016;**10**(2):318-324

[11] Sun Y, Fei T, Pohl N. A high-resolution framework for range-doppler frequency estimation in automotive radar systems. IEEE Sensors Journal. 2019;**19**(23):11346-11358

[12] Rohling H. Radar cfar thresholding in clutter and multiple target situations. IEEE Transactions on Aerospace and Electronic Systems, AES. 1983;**19**(4): 608-621

[13] Bourdoux A, Parashar K, Bauduin M. Phenomenology of mutual interference of fmcw and pmcw automotive radars. In: 2017 IEEE Radar Conference (RadarConf). Seattle, WA, USA: IEEE; 2017. pp. 1709-1714

[14] Kahlert M, Fei T, Tebruegge C, Gardill M. An improved stepped-frequency pmcw waveform for automotive radar applications. In: 2024 15th German Microwave Conference (GeMiC). Duisburg, Germany: IEEE; 2024. pp. 193-196

[15] Braun KM. OFDM Radar Algorithms in Mobile Communication Networks [Thesis]. Karlsruhe: Karlsruher Institut für Technologie (KIT); 2014

[16] Uysal F. Phase-coded fmcw automotive radar: System design and interference mitigation. IEEE

Transactions on Vehicular Technology. 2020;**69**(1):270-281

[17] Jeannin M, Lang O, Khalid FB, Nugraha D, Achatz S, Eryildirim A, et al. Modeling and removing doppler division multiplexing spurs in automotive mimo radar. IEEE Sensors Journal. 2023;**23**(2): 1389-1396

[18] Engels F, Heidenreich P, Wintermantel M, Stäcker L, Al Kadi M, Zoubir AM. Automotive radar signal processing: Research directions and practical challenges. IEEE Journal of Selected Topics in Signal Processing. 2021;**15**(4):865-878

[19] Zheng R, Sun S, Kuo W, Abatzoglou T, Markel M. 4d automotive radar exploiting sparse array optimization and compressive sensing. In: 2023 57th Asilomar Conference on Signals, Systems, and Computers. Pacific Grove, CA, USA: IEEE; 2023. pp. 1-5

[20] Zheng R, Sun S, Liu H, Teresa W. Deep-neural-network-enabled vehicle detection using high-resolution automotive radar imaging. IEEE Transactions on Aerospace and Electronic Systems. 2023;**59**(5): 4815-4830

[21] Schüßler C, Hoffmann M, Ullmann I, Ebelt R, Vossiek M. Deep learning based image enhancement for automotive radar trained with an advanced virtual sensor. IEEE Access. 2022;**10**:40419-40431

[22] Grimm C, Fei T, Warsitz E, Farhoud R, Breddermann T, Haeb-Umbach R. Warping of radar data into camera image for cross-modal supervision in automotive applications. IEEE Transactions on Vehicular Technology. 2022;**71**(9):9435-9449

[23] Alland S, Stark W, Ali M, Hegde M. Interference in automotive radar systems: Characteristics, mitigation techniques, and current and future research. IEEE Signal Processing Magazine. 2019;**36**(5):45-59

[24] Kunert IM. Mosarim: More Safety for all by Radar Interference Mitigation. Technical Report, European Commission; The Contract no.: 248231. 2012. Available from: https://cordis.europa.eu/docs/projects/cnect/1/248231/080/deliverables/001-MOSARIM Deliverable53V10.pdf

[25] Ossowska A, Sit L, Manchala S, Vogler T, Vanova J, Hejtmanek J, et al. Imiko-radar: Interference measurements of today's automotive radar sensors. In: 2021 21st International Radar Symposium (IRS). Berlin, Germany: IEEE; 2021. pp. 1-6

[26] Fei T, Guang H, Sun Y, Grimm C, Warsitz E. An efficient sparse sensing based interference mitigation approach for automotive radar. In: 2020 17th European Radar Conference (EuRAD). Utrecht, Netherlands: IEEE; 2021. pp. 274-277

[27] Liu S, Zhang Z, Fei T, Gong Z, Kou L, Shan D, et al. Design and performance validation of cwt-mca based interference mitigation for automotive radars. Digital Signal Processing. 2024;**153**:104644

Chapter 3

Conceptualizing Automobile Emission Control System: A Key to Achieving Sustainable Automotive Repair Practices and Reducing Vehicle Air Pollution in Sub-Saharan Africa

Zubairu Abubakar and Abubakar Sadiq Adamu

Abstract

This chapter conceptualizes automobile emission control system for service technicians in Sub-Saharan Africa with a view to improving the maintenance of internal combustion engine (ICE) vehicles that use fossil fuel. This is to make them cleaner and more fuel-efficient, thereby reducing vehicle air pollution in the continent. This aligns with sustainable development goals (SDG 11), "sustainable cities and communities," and SDG (13), "climate action," and also supports national development plans in many countries. These concerns have been the reasons behind many technological developments in automotive vehicles, such as the integration of onboard diagnostics (OBD) systems, electronic fuel injection, engine management systems, and many other emission reduction technologies. However, in Sub-Saharan Africa, the automobile emission control system has not been given the care it deserves. Most automotive services technicians in the continent embark on unprofessional practices in handling emission-related concerns and maintenance of the emission control system, which alters the precision of some components, resulting in more emissions. This chapter, therefore explains in clear terms the automobile emission control system and also highlights relevant competencies for automotive service technicians with a view to reducing vehicle air pollution through effective maintenance of the prevalent ICE vehicles in Sub-Saharan Africa.

Keywords: automobile emission control system, sustainable automotive repairs, vehicle air pollution, sub-Saharan Africa, sustainable development goals, automotive service technicians

1. Introduction

The natural environment faces a lot of problems as a result of human activities, which increase with time. The problems cause major threats that affect the well-being of the environment, human beings, and all other living organisms on the surface of the earth. Environmental problems constitute one of the key challenges of the twenty-first century [1]. They include overpopulation, waste disposal, global warming, and climate change, primarily caused by air pollution. According to the World Health Organization [2], air pollution is now recognized as the single biggest environmental threat to human health and a primary contributor to climate change. Air pollution is the introduction of chemicals, particulate matter, or biological materials into the atmosphere that cause harm or discomfort to humans and other living organisms and can degrade the natural environment [3]. Air pollution comes either from natural sources or from man-made sources, such as the burning of fossil fuels for powering stationary engines and mobile transportation.

Transport sector contributes immensely to increasing the level of pollutants in many cities. Transportation is the main source of pollutants in many urban areas across the globe [4]. Mobile transportation, according to Ogola [5], contributes to greenhouse gas emissions and many air pollution emissions through combustion of fossil fuels, mainly from diesel and petrol consumption for mobile transportation. This involves the use of motor vehicles powered by internal combustion engines. The engines produce power for use through combustion of fuel and air, resulting in the generation of pollutants from fuel fumes and byproducts of combustion known as automobile emissions.

Automobile emission is the release of toxic gases and other substances as a byproduct of combustion or fuel fumes from automobile sources. They include Carbon monoxide (CO), hydrocarbon (HC), Nitrogen Oxides (NO and NO_2), Particulate Matter (PM), and carbon dioxide. These pollutants extensively pose a threat to human health and the environment. For instance, carbon monoxide is very poisonous to human health; hydrocarbons react with nitrogen oxide to form ground-level ozone (O_3), which is a major component of smog. Nitrogen oxides contribute to smog and acid rain, causing irritation to human mucus membranes. Carbon dioxide is the major cause of Green House Gases (GHGs) emissions and climate change. In this regard, Yang et al. [4] regarded automobile emissions as a threat to public health, reducing air quality and significantly causing climate change and environmental degradation [4, 6, 7]. The adverse effects of these noxious emissions and immediate needs to control them forced the development of federal emission standard regulations and policies at the global level, such as the Clean Air Act of 1970 and 1977.

To meet these standards, manufacturers incorporated many technological developments into vehicle designs, which include electronically controlled fuel feed and ignition systems, engine management systems, and onboard diagnostics systems. Others are emission reduction technologies such as three-way catalytic converters, evaporative emission control systems, positive crankcase ventilation, exhaust gas recirculation, air injection systems, and diesel particulate filters peculiar to diesel engines. These components of the emission control system serve three main purposes if functioning properly; to promote more complete combustion in order to reduce the byproducts of combustion, to redirect unburnt hydrocarbon into the engine for proper combustion, and to create more area for oxidation and combustion to take place [8].

However, despite these efforts by automobile manufacturers, vehicles in Sub-Saharan Africa generate considerable emissions beyond acceptable thresholds. Motor vehicles in developing countries cause serious air pollution due to poor mechanical conditions and poor functioning emission control systems [9, 10]. The release of toxic gases increases due to the continuous growth of the vehicle population coupled with poor mechanical conditions and less emphasis on control of vehicular emissions. In Nigeria, for instance, the automotive service technicians are more familiar with repairs of old vehicles with poorly functioning emission control systems. In the case of new model vehicles, the technicians avoid emission-related concerns, especially when a fault code is set by the vehicle's onboard computer [10]. Other technicians intentionally disable the functionality of some components due to fear of the system's self-reporting capability. It is a known fact that when a component of an automobile emission control system is not working properly or is altered, the result is more emissions [11].

These unsustainable practices by automotive service technicians in Sub-Saharan Africa continue to pose great challenges to the environment and its occupants and thereby specifically hinder the achievement of sustainable development goals (SDG 11 and 13), "sustainable cities and communities," and "climate" [12]. While many developed and some developing countries are set to achieve net zero emissions by reducing their carbon footprint through vehicle electrification, the use of ICE vehicles with poorly functioning emission control devices is still getting prevalence in Sub-Saharan Africa. The worst scenario is that automotive service technicians who are charged with automotive vehicle maintenance take part in tampering with the emission control system components simply due to ignorance. In view of the above, this chapter will explain in clear terms what an automobile emission control system is all about, how the components work, and finally highlight relevant competencies for automotive service technicians in Sub-Saharan Africa.

2. Concept of automobile emission

Automobile emission is the release of toxic gases as a byproduct of combustion and fuel fume evaporation from automobile sources. The emitted gases form part of other pollutants that present a major threat to human health and the environment. According to Raji et al. [13], automobile emission is a serious environmental health problem in the world today. The automobile emissions include carbon monoxide (CO), hydrocarbon (HC), nitrogen oxide (NO_X), particulate matter (PM), and carbon dioxide (CO_2). The pollutants can be classified into those that are directly harmful to human health and the environment and those that are toxic when combined with other compounds. These emissions from automobiles threaten public health, reduce air quality, and significantly cause climate change and environmental degradation [4]. Thus, it is imperative to identify the sources of the emissions in motor vehicles and how they are controlled.

Automobile emissions can be a result of incomplete combustion in the engine or due to fuel fume evaporation. The power to move a car comes from burning fuel and air inside an engine, which in turn generates power to propel the vehicle. However, the combustion process is not 100% efficient; it can be affected by inappropriate mixture formation, leaks in the cylinder, and very high temperatures during combustion, as well as faults in the fuel or ignition system. Any of the above conditions can considerably affect the combustion process efficiency and also lead to the formation of toxic gases. Other factors are mechanical condition of the vehicle, load consideration, and driving condition. According to Ibekwe et al. [14], incomplete combustion of air and fuel in the engine cylinder results in the development of harmful emissions. Additionally, fuel fumes can also evaporate from the fuel tank, carburetor, and engine crankcase, getting their way out to pollute the environment.

3. Sources of automobile emissions

Automobile emissions usually come from three main sources; these are fuel tanks and carburetors, which vaporize 15% of HC emission, crankcase emits 20% of the HC, and tailpipes, which emit 60% [8]. Other toxic gases emitted through the tailpipe are CO and NO_X, and SO_2 and PM in diesel engines. Emission of CO_2 also follows suit through the tailpipe; however, CO_2 is not directly toxic but significantly causes climate change.

3.1 Fuel tank and carburetor

Fuel tank is a storage container for combustible fuel, usually made from steel, aluminum, or plastics. It is located somewhere between the rear and front of a vehicle such that it stays safe in the event of a collision. The tank has a sealed gas filler cap provided to allow for easy refueling and to control fuel vapor from escaping to the atmosphere. At the neck of the tank, a specially designated valve is provided by mechanical means to open during refueling and closes immediately when the fuel supply hose is withdrawn. By doing so, the fuel and fuel vapor remain in the tank for use. Vapors from the stored fuel keep the tank under pressure, which must be vented to avoid the tank from exploding or bursting. Fuel, especially petrol, is known to release hydrocarbon vapors, which, according to Bonnick and Newbold [15], must not be allowed to escape into the atmosphere because of its harmful effects on human health and the environment. Apart from the fuel tank, a carburetor found in old model vehicles is also a source of hydrocarbon emissions in Sub-Saharan Africa because these types of vehicles are still found on the roads. The carburetor meters air and fuel in a predetermined ratio for use during combustion. In the process, fuel fumes vaporize, contributing to atmospheric pollution. To control the hydrocarbon emission from the tank and carburetor, an Evaporative Emission Control (EVAP) system technology is used (**Figure 1**).

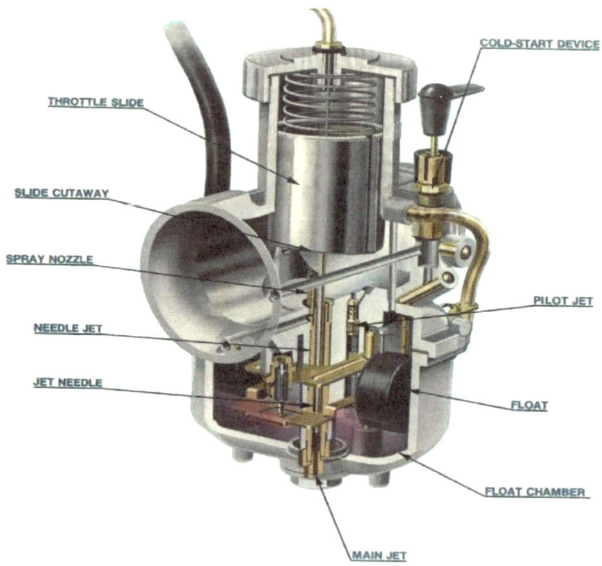

Figure 1.
A carburetor.

3.2 Crankcase

Another source of automobile emission is the crankcase, a portion between the engine block and oil sump. It houses the engine crankshaft and is most often integrated as a single unit with the engine block. Exhaust from the combustion chamber and unburned hydrocarbons find their way to the crankcase, usually through the cylinder walls. These emissions, also known as "blowby," mix with engine oil to form sludge-like compounds that sediment in the engine compartment, affecting parts lubrication and engine efficiency. Initially, provision was made to vent these gases outside the engine, yet creating the worst scenario by polluting the atmospheric air. Now with the introduction of the Positive Crankcase Ventilation (PCV) system, the problem of emissions from the engine crankcase is addressed (**Figure 2**).

3.3 Tailpipe

Major automobile emissions come from the combustion of fuel and air in the combustion chamber during the normal combustion process. Exhaust gases produced are channeled out through the exhaust manifold and then to the tailpipe, hence termed

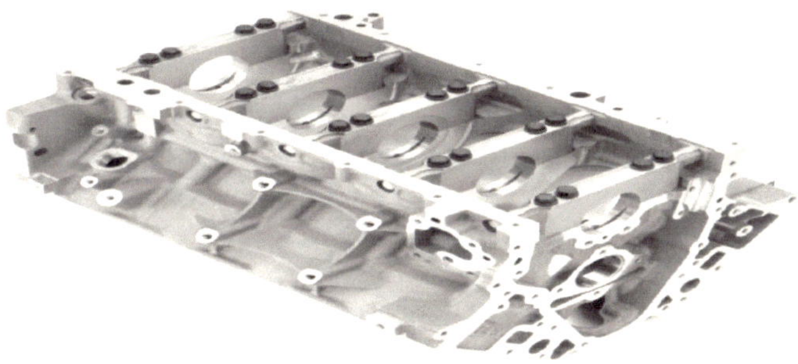

Figure 2.
Engine crankcase.

Figure 3.
Tailpipe.

tailpipe emissions. Categories of the emissions released through the tailpipe include hydrocarbons, carbon monoxide, oxides of nitrogen, carbon dioxide, as well as sulfur oxide and particulate matter emissions in diesel engines. This constitutes a higher percentage of the overall emissions released by automobiles. For instance, the tailpipe emits 60% of hydrocarbons, while the crankcase and fuel tank emit only 20 and 15%, respectively. Thus, a number of systems known as post-combustion exhaust emission control technology are put in place to control the tailpipe emissions. A typical example of such a technology is the catalytic converter (**Figure 3**).

4. Automobile emission control technologies

Automobile manufacturers depend solely on technology to reduce pre- and post-combustion emissions from motor vehicles [16]. This gave birth to the automobile emission control system and the incorporation of emission reduction technologies in motor vehicles. These technologies include Evaporative Emission control (EVAP) system to control fuel fumes evaporating from the fuel tank and carburetor; Positive Crankcase Ventilation (PCV) system to control "blowby" gas emissions from the crankcase; and Exhaust Gas Recirculation (EGR) to lower peak combustion temperature so as to reduce Nitrogen oxides (NO_X) emissions. Additionally, a device known as a catalytic converter controls HC, CO, and NO_X emissions. To control Particulate Matter (PM) emissions in diesel engines, a diesel particulate filter is used. Other related systems are onboard diagnostic system generation II, electronic fuel injection, and engine management system. All these systems contribute in one way or another toward achieving desired automobile emission reduction. According to Halderman [8] and Bicam [11], if any of these systems is missing, altered, or tampered with, the vehicle will produce more emissions. Thus, it is imperative to explain at a layman's level how these emission reduction technologies work and, to some extent, enumerate maintenance clues for automotive service technicians in Sub-Saharan Africa.

4.1 Evaporative emission control (EVAP) system

Evaporative Emission Control (EVAP) system controls hydrocarbon emissions from the fuel tank and related components such as the carburetor in old model vehicles. Special provision is made by the EVAP system to utilize evaporating fuel vapors that put the tank under pressure instead of venting it to the atmospheric air. The system traps, stores, and directs the fuel vapors to the intake air, where it is burned along with fuel in the combustion chamber. In vehicles equipped with a modern EVAP system, a specially designed filler tank cap is used. The cap utilizes a pressure-vacuum relief valve that opens and closes on the basis of pressure and vacuum action in the tank in relation to a predetermined threshold. When the valve opens, the fuel vapor passes to a very important component of the system known as the charcoal canister, which is connected via a hose to the throttle body or air cleaner, where it mixes with air-fuel charges and finally to the engine cylinder for burning. In a typical EVAP system, the computer controls the operation of the system in relation to the manifold vacuum (**Figure 4**) [8].

4.2 Positive crankcase ventilation system (PCV)

To control the emission of leaked combustion gases in the engine crankcase, a Positive Crankcase Ventilation (PCV) system technology is used. The system utilizes "blow by"

DOI: http://dx.doi.org/10.5772/intechopen.1009355

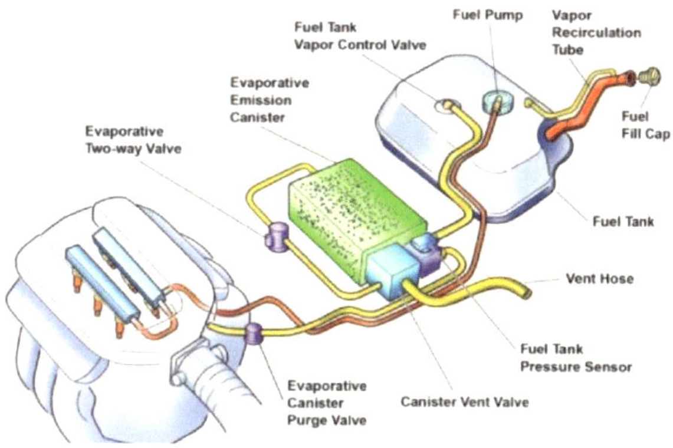

Figure 4.
EVAP system.

gases in the engine crankcase by routing them back to the intake system and to the combustion chamber, where they undergo another combustion process as a means to ventilate the crankcase and reduce crankcase vapor emissions. The PCV system consists of a valve or calibrated orifice, an inlet air filter, and connecting hoses. The PCV valve regulates the flow of air from the crankcase based on driving conditions and speeds [8]. In the event of high manifold vacuum, such as during idle, when the air/fuel mixture is very critical, just a little of the vapors are allowed into the intake system, but at high speed, when the mixture is less critical and the pressures in the engine are greater, more of the vapors are allowed into the intake system. Faults in the PCV system include restrictions or blockages in the systems, which should be checked before attempting major repairs (**Figure 5**).

4.3 Exhaust gas recirculation system (EGR)

Exhaust gas recirculation system is used to control nitrogen oxide emission by lowering peak combustion temperature. Oxygen in the air used for combustion

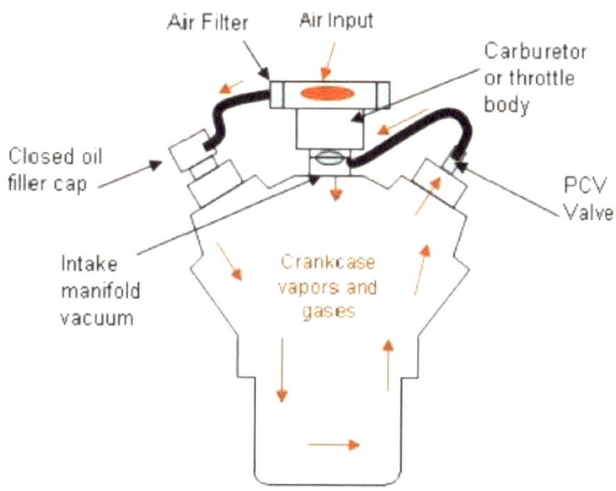

Figure 5.
PVC system.

contains abundant nitrogen. And when nitrogen is exposed to a high temperature of about 1800 K, an oxide of nitrogen is formed [15]. The EGR system prevents this condition by keeping the combustion chamber temperature below the critical point. It achieves this by directing a small portion of the exhaust gases to the intake manifold. This dilutes and reduces the amount of active intake charges into the combustion chamber, which lowers peak combustion temperature, a condition for the creation of the NO_X emission. The link between the EGR system and intake manifold is through a valve known as the EGR valve, which can be mechanically or electronically operated. The electronically controlled EGR valve provides a means through which the exhaust gases are routed to the intake manifold. This means that the condition for the opening and closing of the EGR valve is managed by the electronic control unit of the engine management system. Maintenance for the EGR system involves cleaning and replacement of the EGR valve, checking and repairing of vacuum leaks, and using of OBD II scan tools to read and interpret EGR-related diagnostic codes (**Figure 6**).

4.4 Secondary air injection system (SAI)

The function of the secondary air-injection (SAI) system is to provide the required air needed for the oxidizing process either at the exhaust manifold or inside the catalytic converter. Since no internal combustion engine is 100% efficient, there will always be some unburned fuel in the exhaust that increases hydrocarbon emissions; thus, the air injection system eliminates the source of the emissions. Combustion requires fuel, oxygen, and heat. Without any one of the

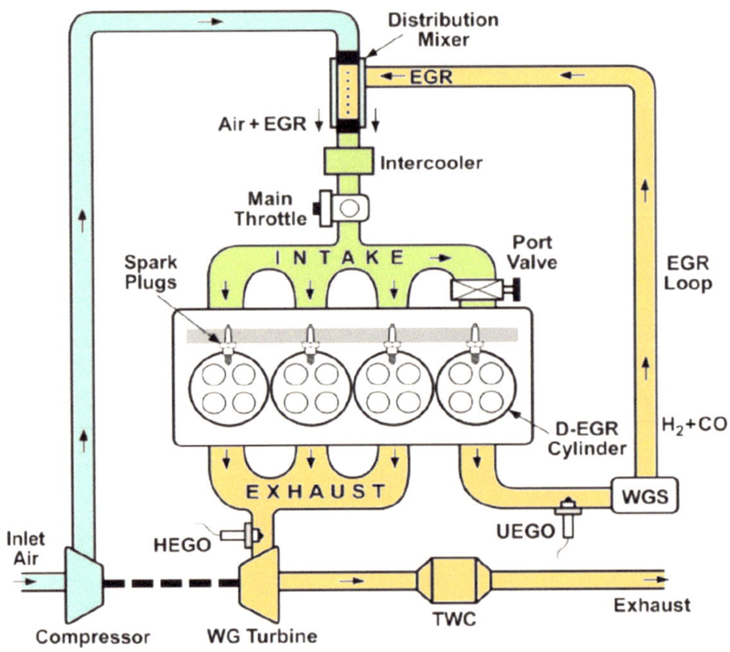

Figure 6.
EGR system.

three, combustion cannot occur. Inside the exhaust manifold, there is sufficient heat to support combustion; if we introduce some oxygen, then any unburned fuel will ignite. This combustion will not produce any power, but it will reduce excessive hydrocarbon emissions. Unlike in the combustion chamber, this combustion is uncontrolled, so if the fuel content of the exhaust is excessive, explosions that sound like popping will occur. According to Halderman [8], the major maintenance that is required is a careful inspection of all SAI system parts and, more importantly, the air pump drive belt for wear or damage.

4.5 Catalytic converter

Catalytic converter is a post-combustion emission control device that controls the emission of HC, CO, and NO_X through oxidation and reduction reactions.

The converter provides an additional area for oxidation or combustion to occur. It looks like a muffler and is located in the exhaust system ahead of the muffler. The catalytic converter converts exhaust gases into water vapor (H_2O) and carbon dioxide (CO_2). A three-way catalytic converter, usually found in modern vehicles, performs the following three major tasks:

a. To reduce nitrogen oxides to nitrogen and oxygen through a reduction reaction. Hence, the equation, $2NO_x$ - $N_2 + O_2$

b. To oxide the unburnt non-methane hydrocarbon to carbon dioxide and water. Hence, $C_xH_y + nO_2$ - $xCO_2 + mH_2O$

c. To oxidize carbon monoxide to carbon dioxide

Thus, $2CO + O_2 = 2CO_2$.

As the converter works to clean the exhaust, it develops heat. The dirtier the exhaust, the harder the converter works and the more heat that is developed. In some cases, the converter can be seen to glow from excessive heat. If the converter works this hard to clean a dirty exhaust, it will *destroy itself*. Maintenance of the catalytic converter includes various tests and replacements where necessary (**Figure 7**).

Emission reduction technologies so far explained controlling exhaust emissions and hydrocarbon vapors. **Table 1**, adapted from Halderman [8], presents a summary of exhaust gases, their causes, and diagnosis clues.

Figure 7.
Catalytic converter.

Emitted gas	Cause	Diagnosis clue
High HC	Engine misfire or incomplete burning of fuel	i. Ignition system fault ii. Lean misfire iii. Too low an engine temperature
High CO	Rich condition	i. Leaking fuel injectors or fuel-pressure regulator ii. Clogged air filter or PCV system iii. Excessive fuel pressure
High HC and CO	Excessively rich condition	i. All items included under high CO ii. Fouled spark plugs causing a misfire to occur iii. Possible non-operating catalytic converter
High NO_X	Excessive combustion chamber temperature	i. Non-operating EGR valve ii. Clogged EGR passages iii. Engine operating temperature too high due to cooling system restriction, worn water pump impeller, or other faults in the cooling system iv. Lean air-fuel mixture v. High compression caused by excessive carbon buildup in the cylinders

Table 1.
Exhaust gas diagnosis summary guide.

5. Automobile emission control systems in Sub-Saharan Africa: Challenges and opportunities

Despite the fact that ICE vehicles are still the primary source of many emissions, their use is getting more prevalent in many Sub-Saharan African countries, subjecting the countries to more emission problems. The rapid rise in vehicle ownership has outpaced the implementation of stringent emission control regulations, leading to a significant increase in air pollution and associated health problems. In these regions, vehicles, often older models, lack functional emission control systems, resulting in the emission of harmful pollutants such as hydrocarbons (HCs), soots, carbon monoxide (CO), nitrogen oxides (NOx), and particulate matter (PM) [2]. According to a report from the Institute for Transportation and Development Policy [17], air pollution is one of the leading causes of respiratory and cardiovascular diseases globally, and Sub-Saharan African countries are significantly affected due to the combination of high vehicle emissions and industrial pollution.

Economic challenges in these countries often result in a higher concentration of older, secondhand vehicles, many of which are imported from developed nations. In Nigeria, for instance, these older, secondhand vehicles, referred to as Tokunbo, in most cases, are not equipped with the latest emission control technologies or may be in poor condition, further exacerbating pollution. In Kenya, a study by the Institute for Transportation and Development Policy (ITDP) found that more than 60% of the vehicles in Nairobi were more than 10 years old, many of which lacked adequate emission controls [17]. The lack of proper maintenance and technical knowledge among mechanics is another key issue due to a lack of necessary tools, expertise, and knowledge of parts specifications to properly maintain or make appropriate part replacements in the emission control systems. A survey conducted in Uganda found that many mechanics were unaware of the environmental impact of poorly maintained vehicles and lacked the skills to repair advanced emission control components such as catalytic converters and oxygen sensors [18]. In another study conducted in Nigeria [19] indicated that the knowledge of roadside mechanics about automobile waste management and disposal in northeast Nigeria is relatively low, especially their knowledge of the environmental impacts associated with automobile waste, both solid, liquid, and gas. This lack of knowledge, expertise, and infrastructure undermines the effectiveness of existing emission control systems, making it difficult for Sub-Saharan Africa to reduce vehicle emissions and improve air quality. This calls for a conceptual understanding of automobile emission control systems and, in general, the need for sustainable automotive repair practices in Sub-Saharan Africa.

6. Achieving sustainable automotive repair practices in Sub-Saharan Africa

Sustainable development goals (SDG 11 and 13) focused on reducing climate change problems and ensuring sustainable cities and communities across the globe, respectively [12]. These are associated with addressing the causes and sources of environmental air pollution. Vehicle emissions have been the cause of increased greenhouse gases (GHGs) and general environmental air pollution. Thus, the implementation of sustainable automotive repair practices is crucial for achieving "climate action" and "sustainable cities and communities," especially in Sub-Saharan Africa, which is always at the receiving end when it comes to environmental pollution attached to end-of-use vehicles. Sustainable automotive repair practices focus on ensuring that vehicles are well-maintained and their emission control systems are functioning optimally. However, challenges such as limited access to spare parts, outdated repair techniques, and low levels of technical training among service technicians hinder the adoption of such practices.

In many countries, the repair industry is largely informal, with many mechanics relying on outdated tools and methods to repair vehicles. For instance, in Nigeria (the largest African economy), technical training for mechanics is heavily, if not entirely, informal, with the traditional apprentice method as the major mode of training. So also in other Sub-Saharan African countries like Ghana, Niger, Cameroon, and many others, the training of automotive service technicians is often informal and varies significantly across regions. In this training model, young people, often from low-income families, join established mechanic workshops as apprentices. The apprenticeship typically lasts anywhere from 3 to 5 years, during which time apprentices learn through hands-on experience, working alongside skilled mechanics who serve as mentors. The focus is usually on practical skills with no emphasis on theoretical knowledge that underlines operational principles and effective functioning of the vehicle components in relation to predetermined standards set by the vehicle manufacturers. The apprentices are mainly coached to repair basic mechanical issues, such as engine overhauls, brake systems, and basic electrical system faults, but are often not exposed to more modern and specialized trends in vehicle repairs, such as emission control systems maintenance, electronic engine diagnosis, and computerized exhaust gas analysis, as well as repair and maintenance of hybrid (HEV) and electric vehicles (BEV). The prevalence of automotive service technicians who lack requisite competencies and technical support leads to the improper handling of emission control systems, and the continued use of in-service vehicles with poorly functioning emission control devices is attributed to the conventional apprenticeship mode of training in Sub-Saharan Africa.

One promising approach that may be of immense benefit in Sub-Saharan Africa is the adoption of "green" repair initiatives. This approach will emphasize good repair and maintenance of emission control system components, appropriate use of diagnostic tools to address emission-related malfunctions, as well as a vehicle emission test and certification program for local mechanics. Additionally, the program will ensure the use of eco-friendly materials, energy-efficient technologies, and the recycling of vehicle components in all automotive repair shops. In Brazil, the "Green Mechanic" program has also been successful. The program trained thousands of mechanics in sustainable automotive repair practices, focusing on the maintenance of emission control systems and the use of environmentally friendly parts [20]. This initiative has been instrumental in reducing vehicle emissions by ensuring that repair shops follow best practices in emission system maintenance. Similarly, the African Development Bank (AfDB) has supported training programs in several African countries, aiming to improve the capacity of local technicians to repair and maintain modern emission control technologies. Through partnerships with local governments, international NGOs, and private sector stakeholders, these programs can help build a workforce capable of supporting sustainable automotive repair practices, ultimately reducing emissions and improving air quality.

Moreover, regulatory frameworks that incentivize the proper maintenance of vehicles and the adoption of greener technologies can accelerate progress. In South Africa, for example, the government introduced a vehicle emissions test requirement for all vehicles, coupled with penalties for non-compliance and incentives for compliance. This regulation encourages vehicle owners to maintain their cars and trucks in line with international emission standards [11]. In addition, the introduction of policies that support the importation of cleaner, newer vehicles and the promotion of green technology adoption, such as electric vehicles (EVs) and hybrid vehicles, are also crucial for long-term sustainability. By improving the competency of automotive services technicians through expanding training opportunities and creating a regulatory framework that supports sustainable automotive repair practices and clean vehicle adoption, Sub-Saharan Africa can make substantial improvements in reducing vehicle-related air pollution and ensure the realization of sustainable cities and communities across the continent.

7. Competency needs of automotive service technicians for effective maintenance of automobile emission control system in Sub-Saharan Africa

Automobile emission control system is a complex system that makes use of electrical/electronic, mechanical, sensor, and computer technologies to accomplish desired emission reduction. The system encompasses a reasonable number of technologies due to increasing demands from legislators to reduce vehicle emissions [15]. With this effort made, the emission control system, if working properly, will guarantee better performance, good fuel economy, and environmental well-being. On the contrary, poor maintenance of the system deters the achievement of the performance characteristics and seriously affects the environment's well-being. Therefore, the task of properly maintaining and making good repairs on automobile emission control systems presents automotive service technicians in Sub-Saharan Africa with an entirely new set of challenges. Hence, developing relevant competencies to take proper care of such an important system is a must for the automotive service technicians.

Component	Remark
Knowledge	
a. Demonstrate knowledge of automobile emissions and their negative effects on health and environment b. Know the rationale and function of automobile emission control system and components c. Understand the function of catalytic converter d. Understand the function of positive crankcase ventilation system e. Understand the function of exhaust gas recirculation (EGR) system f. Understand the function of secondary air injection system g. Understand the function of evaporative emission control system (EVAP) h. Understand the function of diesel particulate matter filters and selective catalytic reduction technology in diesel engines i. Demonstrate sound knowledge of On-Board Diagnostics (OBD II) system generation two in relation to control of automobile emissions j. Describe the procedures for using OBD II scan tool to read, interpret, and erase Diagnostic Trouble Codes (DTCs) k. Know the various emission regulation standards and relevant government agencies for control of vehicle emissions in their various countries	Relevant knowledge learning outcomes on automobile emission control systems for automotive service technicians
Skills	
a. Carry out basic inspection and visual checks for automobile emission control system and related components b. Make use of appropriate equipment to test various components of automobile emission control system c. Carry out various tests on catalytic converter and make appropriate replacements of the converter where the need arises d. Rectify fault in the exhaust gas recirculation system to include removing, cleaning, and replacing faulty EGR valve e. Rectify fault in evaporative emission control system to include checking purge valve and replacing charcoal canister filter f. Clean and replace engine coolant temperature sensor, mass air flow sensor, and oxygen sensor g. Diagnose and rectify faults relating to fuel injection system, which include low fuel pressure and service/replacement of fuel injectors h. Inspect, adjust, and replace air pump drive belt and air filter i. Carryout engine performance test and interpret the result in relation to emission variability j. Remove, service, and install a new spark plug to specification k. Carry out computerized exhaust gas analysis and tailpipe emission tests using appropriate equipment l. Use OBD II live scan mode to check proper operation of engine coolant temperature sensor, mass air flow sensor, and oxygen sensor m. Check coolant temperature resistance using a digital multimeter n. Use OBD II scan tool to read, interpret, and erase diagnostic trouble codes (DTC)	Relevant skill learning outcomes on automobile emission control system for automotive service technicians

Component	Remark
Attitude	
a. Observe safety rules while working on automobile emission control system b. Exhibit environmental consciousness while working on automobile emission control system and general vehicle maintenance c. Appreciate relevant rules governing control of automobile emission d. Develop inquisitive mind toward automobile emission control system e. Take part in campaigning for proper care and good handling of emission control systems among technicians and motorists	Relevant attitude learning outcomes on automobile emission control systems for automotive service technicians

Table 2.
Competency framework on automobile emission control system for automotive service technicians in Sub-Saharan Africa.

Competency refers to requisite knowledge, appropriate skills, and the right attitude to carry out one's job to desired specifications. In the realm of automobile emission control systems, a service technician needs to have sound knowledge of automobile emissions, their causes, and the rationale for controlling them. Developing a conceptual understanding of methods of controlling automobile emissions, technologies used to control the emissions, and how they function is also of utmost importance. Additionally, automotive service technicians need to develop appropriate skills for the repair and maintenance of the automobile emission control system as well as how to carry out relevant tests of the components and exhaust emissions using appropriate equipment. Finally, the technicians should demonstrate the right attitude and environmental consciousness in handling emission-related malfunctions and general vehicle maintenance. Thus, to address these challenges, programs, and frameworks that focus on upgrading the status of local mechanics toward sustainable automotive repair technologies are essential. **Table 2**, adapted from Abubakar [10], presents a framework of competencies related to the maintenance of automobile emission control systems for automotive service technicians in Sub-Saharan Africa.

8. Conclusion

Conceptual understanding of automobile emission control systems is a prerequisite to effective management of automobile emission control systems and a roadmap to sustainable automotive repair practices among automotive service technicians in Sub-Saharan Africa. This entails good management of emission control system components as well as effective maintenance/repair of general vehicle systems and subsystems that may affect drivability. Therefore, by equipping automotive service technicians with the requisite knowledge, skills, and the right attitude for effective handling of automobile emission control systems and adopting relevant regulatory practices, Sub-Saharan Africa can reduce vehicle-related air pollution significantly.

Author details

Zubairu Abubakar* and Abubakar Sadiq Adamu
Department of Mechanical Technology Education, Abubakar Tafawa Balewa University, Bauchi, Nigeria

*Address all correspondence to: zabubakar2@atbu.edu.ng

IntechOpen

© 2025 The Author(s). Licensee IntechOpen. This chapter is distributed under the terms of the Creative Commons Attribution License (http://creativecommons.org/licenses/by/4.0), which permits unrestricted use, distribution, and reproduction in any medium, provided the original work is properly cited.

References

[1] Avogbe PH, Ayi-Fanou L, Cachon BI, Chebi N, Debende A, Dewaele D. Haematological changes among Beninese motor bike taxi drivers exposed to benzene by urban air pollution. African Journal of Environmental Science and Technology. 2011;7:464-470

[2] World Health Organization. Global Air Quality Guidelines: Particulate Matter ($PM_{2.5}$ and PM_{10}), Ozone, Nitrogen Dioxide, Sulfur Dioxide and Carbon Monoxide. Geneva: WHO Publications; 2021

[3] Anukam LC. Air quality monitoring and management challenges. The Interne. 2015. Available from: https://cdn.cseindia.org

[4] Yang Z, Wang H, Shao Z, Muncrief R. Review of Beijing's Comprehensive Motor Vehicle Emission Control Programs. [Internet] Beijing. International Council on Clean Transportation; 2015. Available from: https://theicct.orgcommunications@theicct.org

[5] Ogola SJ. Environment, Climate Change and the Green Economy. Green Economy in the Transport Sector: A Case Study of Limpopo Province. South Africa: Springer; 2021. DOI: 10.1007/978-3-030-86178-0

[6] Srivastava V, Schaub J, Pischinger S. Model-based closed-loop control strategies for flex-fuel capability. Applied Energy. 2023;**350**:121795

[7] Abbas W, Yan M. Performance investigation of solar assisted desiccant integrated Maisotsenko cycle cooler in subtropical climate conditions. Case Studies in Thermal Engineering. 2023;**44**:102864

[8] Halderman JD. Automotive Technology: Principle, Diagnosis and Service. New Jersey: Pearson Education Inc; 2012

[9] Nkwocha AC, Ekeke IC, Kamalu CIO, Kamen FL, Uzondu FN. Environmental assessment of vehicular emission in port-Harcourt city, Nigeria. International Journal of Environment, Agriculture and Biotechnology (IJEAB). 2017;**2**:906-911. DOI: 10.22161/ijeab/2.2.44/

[10] Abubakar Z. Development of a framework on automobile emission control system for inclusion into Colleges of Education Curriculum of Automobile Technology in Nigeria [thesis]. Yola: Modibbo Adama University; 2025

[11] Bicam PB. Technology Innovation in Green Transport: Green Economy in the Transport Sector: A Case Study of Limpopo Province. South Africa: Springer; 2021. DOI: 10.1007/978-3-030-86178-0

[12] Department of Economic and Social Affairs. Sustainable Development. [Internet]. United Nations: United Nations Department of Economic & Social Affairs; 2024. Available from: https://www.sdgs.un.org.goals

[13] Raji WA, Jimoda LA, Odobor JK, Popoola AO. Assessment of vehicular-induced emissions in some selected areas in Benin City, Edo state, Nigeria. Journal of Applied Science and Environmental Management. 2021;**25**:1535-1539. Available from: https://www.bioline.org.br/ja/

[14] Ibekwe BE, Nwobodo NNH, Bendor SA. Evaluation of current technologies to curb air pollution in Nigeria. Advance Journal of Science,

Engineering and Technology. 2019;**4**:10-23. Available from: https://iaspub.org.uk/AJSET/

[15] Bonnick A, Newbold D. A Practical Approach to Motor Vehicle Engineering and Maintenance. Oxford USA: Elsevier Ltd; 2011

[16] Ayetor GK, Innocent M, Sakey MN, Andoh PY. Vehicle regulations in Africa: Import and new vehicle sales. Transportation Research Interdisciplinary Perspectives. 2021;**10**:20-30

[17] Institute for Transportation and Development Policy (ITDP). Vehicle fleet composition and emissions in Nairobi, Kenya [Internet]. In: Africa; ITDP Report. 6;2018. Available from: www.itdp.org

[18] Kasumba M. Challenges in automotive emission control in Uganda: A survey of mechanic practices. Uganda Journal of Environmental Studies. 2017;**5**:45-58

[19] Inti MM, Adamu AS. Knowledge, attitude and practices of automobile waste management and disposal among road side mechanics in North East, Nigeria for greener environment. IOSR Journal of Environmental Science, Toxicology and Food Technology (IOSR JESTFT). 2024;**18**:1-7

[20] Aro A. Green mechanic program: Transforming the automotive repair industry in Brazil. Environmental Policy and Practice Journal. 2020;**15**:22-35

Chapter 4

Experimental Test and Optimization of Energy Additive for Combustible Fuels to Lower Pollutant Gases

Md Abul Kalam, T.M. Indra Mahlia, Arridina Susan Silitonga, Islam Md Rizwanul Fattah, Md Mofijur Rahman, Mohammad Rashed Mia, Muhammad Ali Ijaz Malik, Nicholas Moulis, Anthony Witts, James Witts and Jonathan Sherlock

Abstract

Environmental concerns, economic growth, and population growth are all contributing to the continuous rise in demand for renewable energy. DXP fuels can be synthesized from renewable resources, aligning with global efforts to reduce dependence on non-renewable energy sources and lower the carbon footprint. The DXP fuel was blended with diesel in three different proportions, and these DXP-diesel blends were tested on a 638 cc diesel engine. On average, it was found that the Brake thermal efficiency (BTE) for 2LDiesel + 1ML DXP and 3LDiesel + 1ML DXP decreased by 10.85 and 9.01% compared to diesel. However, the average BTE for 1LDiesel + 1ML DXP increased by 1.62% compared to diesel. On average, the 2LDiesel + 1ML DXP and 3LDiesel + 1ML DXP demonstrated a remarkable drop in CO_2 emissions of 2.81 and 2.28%, respectively, in comparison with diesel fuel. However, the CO_2 emissions for 1LDiesel + 1ML DXP increased by 0.53% in comparison with diesel. In contrast to diesel fuel, 1LDiesel + 1ML DXP, 2LDiesel + 1ML DXP, and 3LDiesel + 1ML DXP showed an average NO emission reduction of 1.37, 2.74, and 0.80%, respectively. In summary, hydrophilic DXP fuels as alternative fuel additives offer significant engine performance, emission reduction, fuel stability, and environmental sustainability.

Keywords: DXP fuel blends, environmental sustainability, diesel engine, exhaust emissions, engine performance

1. Introduction

Global energy consumption has surged due to rapid population expansion. A progressive reduction in fossil fuel reserves will result in a fuel crisis as the demand for energy in all sectors, like transport, agriculture, and residential, increases [1]. Crude oil currently provides 12.2 billion tons of energy annually for the world. By 2035, this energy use will increase to 1.75 billion tons of crude oil [2]. The ability of oil and gas firms to meet the future energy demands of all sectors is significantly threatened by the depletion of fossil fuels. The use of fossil fuel reserves by oil and gas producers to meet energy demands has resulted in a sharp rise in carbon dioxide emissions (CO_2) worldwide [3]. Eliminating combustion emissions that are directly linked to climate change, global warming, and several diseases is the primary goal of switching from fossil fuels to alternative renewable fuels [4]. High airborne particulate matter concentrations can lead to allergies, cancer, and respiratory conditions like asthma and bronchitis [5]. The transportation industry mainly consumes fossil fuels to meet energy demands. Because the transportation sector uses over half of fossil fuels to meet its energy needs, fossil fuels are in danger of running out. The transportation industry is responsible for around one-third (32%) of the world's overall energy usage. Over the previous 10 years, the transportation sector's global energy demand has grown by 39% [6]. This is a result of the growing need for freight and transportation in emerging nations worldwide. Approximately 75% of the world's passenger vehicle transportation energy usage in 2018 came from road travel [7]. As a result, environmental pollution and energy demand are significant concerns. Fossil fuels, the primary source of greenhouse gas emissions, are mostly used in transportation. Half of greenhouse gas emissions would come from the transportation sector by 2030 [8].

Renewable biofuels will replace fossil fuels to meet the energy needs of the transportation sector and eliminate environmental pollution (carbon emissions). The transportation sector's carbon emissions have the most significant impact on environmental health [9]. The toxic gases released by the burning of petroleum fuels have put human health and environmental problems like climate change and global warming in jeopardy. Researchers are searching for renewable fuels to eliminate combustion emissions due to the growing use of fossil fuels and the detrimental effects of exhaust gases on human health and climate change. The combustion of petroleum fuels has produced toxic gases that have put human health and environmental issues like climate change and global warming in jeopardy. To eliminate combustion emissions, researchers are searching for renewable fuels due to the growing use of fossil fuels and the detrimental effects of exhaust gases on human health and climate change [4]. Exhaust gases such as NOx, CO, minute particulate matter, and HC are produced when fossil fuels are burned in diesel engines, and they pollute the air [10]. The high levels of particle matter in the air can cause serious health problems like cancer and respiratory conditions like asthma and allergies [5]. Reliance on fossil fuels can be replaced by sustainable and renewable fuels to adhere to the Kyoto Protocol. Globally, 46 nations have established national goals to raise the proportion of renewable energy in transportation by 20% by 2020 and 40% by 2030 [6]. The foremost drawback of alternative fuels is increased NOx and CO_2 emissions. Much research has been done on biofuels, but none has been done on DXP fuel blends. Therefore, the current book chapter investigates the impact of local Australian fuel blends (DXP-diesel fuel blends) on engine performance.

2. Experimental methodology

Both diesel and DXP fuels are mixed in such proportion to produce 1LDiesel + 1ML DXP, 2LDiesel + 1ML DXP, and 3LDiesel + 1ML DXP fuel blends. The test fuel properties are displayed in **Table 1**.

2.1 Experimental setup

For the current study, the University of Malaya's test rig with a diesel engine modelled as Yanmar (TF 120 M) was used to ascertain engine performance on distinct fuel samples. The schematic perspective of the experimental diesel engine arrangement is shown in **Figure 1**. Furthermore, **Table 2** displays the diesel engine's technical specifications.

2.1.1 Diesel engine performance

Initially, the CI engine properties were investigated by blending diesel and DXP fuel. A graduated measuring cylinder was hooked to the fuel tank to measure the gasoline flow rate. For every 10 mL of gasoline sample, three readings were taken throughout each test, and their average turned out to be the primary value. Time was recorded using a stopwatch. DASTEP8 software was used to extract the brake torque (BT) and speed. BP denotes the crankshaft's available power output. Brake power output as a function of fuel heat input is known as BTE. The exhaust gas analyzer BOSCH (BEA-350) was utilized to examine the emissions from diesel engines, including HC, CO, and NOx. The measuring precision and method to ascertain exhaust gases are described in **Table 3**. Exhaust gas emissions were measured in parts per million (ppm), which should be converted into a globally recognized unit g/kWh according to SAE J177.

Properties	Units	100% Diesel	100% DXP
Calcium (Ca)	ppm	<1	12
Phosphorus (P)	ppm	<1	7
AN (ASTM D664)	mgKOH/g	0.03	0.27
Water (by KF)	ppm	47	1083
Sulfur by EDXRF (ASTM D4294)	ppm	8	70
Carbon (ASTM D5291)	%	85.62	69.79
Nitrogen (ASTM D5291)	%	<0.7	3.42
calorific value (ASTM240)	MJ/kg	45.64	—
Viscosity (ASTM D445) @40°C	mm^2/s	2.451	1.712
Density (ASTM4052)@15°C	g/ml	0.8302	0.8883
Flashpoint (ASTM D93)	°C	<40	68
Elemental composition (ASTM D5291) (C/H)	%	85.62/13.75	69.79/11.70
Cetane number		52.4	—
Cold filter plugging point		−8	—
Oxidation stability (ASTM D2274)	Mg/100 ml	0.03	—

Table 1.
Fuel properties.

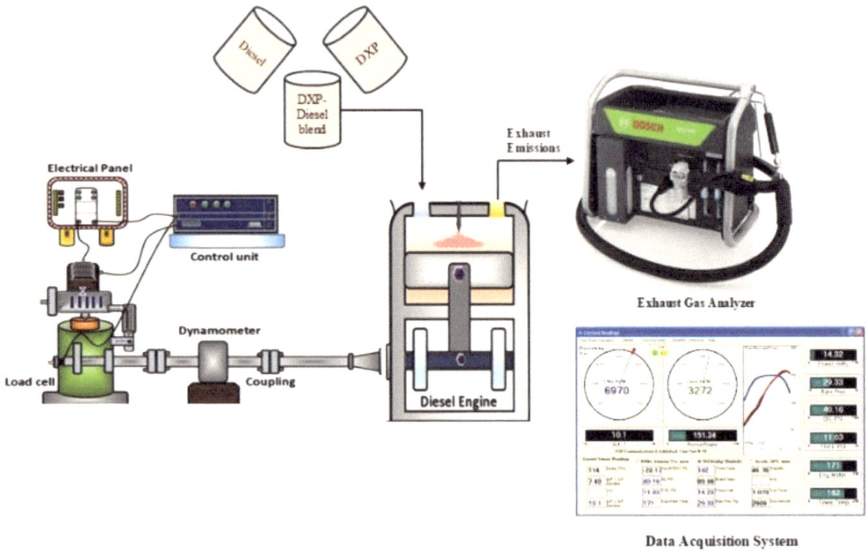

Figure 1.
Test setup schematic.

Engine characteristics	Description	Units
Compression ratio	17.7	
Injection timing	17° BTDC	(deg.)
Displacement	0.638	(L)
Cooling system	Radiator cooling	
No. of cylinders	1	
Maximum engine speed	2400	(rpm)
Maximum power	7.7	(kW)
Aspiration	Radiator cooling	
Injection pressure	200	(kg/cm^2)
Cylinder bore x stroke	92 × 96	(mm)
Connecting rod length	149.5	(mm)
Power take-off position	Flywheel side	

Table 2.
Diesel Engine Characteristics.

To investigate engine characteristics, DXP-diesel fuel blends were first charged into a diesel engine after diesel fuel had been run in the engine to provide stable operating conditions. To remove the tested ternary fuel mix traces, diesel fuel was used to flush the engine after each DXP fuel blend. Diesel, 1LDIESEL + 1ML DXP, 2LDIESEL + 1ML DXP, and 3LDIESEL + 1ML DXP were the fuel samples that were evaluated.

Apparatus	Approach	Parameters	Measurement Range	Accuracy	% Uncertainty	Units
BOSCH BEA 350	Flame ionization detector (FID)	HC	0–9999	±1	±1%	ppm
	Heated vacuum-type chemiluminescence detector (CLD)	NO_x	0–5000	±1 ppm	±1.30%	ppm
	Non-dispersive infrared	CO	0–10	±0.001	±1%	% vol.

Table 3.
Gas analyzer specifications.

3. Diesel engine performance analysis

The engine performance is ascertained based on the impact of Diesel-DXP fuel blends on the following parameters:

3.1 Brake torque

Figure 2 displays the brake torque (BT) findings for each fuel sample at full-throttle openings with varying engine speeds. The comparison between diesel and DXP blended fuel is shown in **Figure 2**. BT values were decreased for all tested fuels from low speeds to high speeds. The maximum BT values of diesel, 1LDiesel + 1ML DXP, and 2LDiesel + 1ML DXP were recorded as 29.64, 29.07, and 28.96 Nm, respectively, at 1500 rpm. However, the maximum torque value for 3LDiesel + 1ML DXP was recorded as 29.75 Nm at 1200 rpm. The minimum BT values for test fuels of 26.12, 26.35, 25.44, and 25.89 Nm were obtained at 2400 rpm for diesel, 1LDiesel + 1ML DXP, 2LDiesel + 1ML DXP, and 3LDiesel + 1ML DXP, respectively. Compared to diesel fuel, the 1LDiesel + 1ML DXP and 2LDiesel + 1ML DXP portrayed a decline in average brake torque values of 0.48 and 1.83%, respectively.

However, the 3LDiesel + 1ML DXP fuel blend showed a 0.24% increase in BT compared with diesel fuel. The 3LDiesel + 1ML DXP fuel blend has a higher BT value because of complete combustion, which produces more energy and raises the average pressure in the engine cylinder, which raises BT and piston force.

Other tested fuels have lower brake torque because the combustion chamber is not adequately filled with fully opened vales during the intake stroke. This reduces volumetric efficiency because there is insufficient time for a sufficient air intake phase, which lowers the combustion chamber pressure.

3.2 Brake power

The brake power of different tested fuels is displayed in **Figure 3**. As the engine speed increased, the brake power progressively increased as well. All tested fuels, including diesel, 1LDiesel + 1ML DXP, 2LDiesel + 1ML DXP, and 3LDiesel + 1ML DXP, had maximum BPs of 7.84, 7.94, 7.62, and 7.74 kW at 2400 rpm, respectively. The minimum BP of 4.45, 4.37, 4.30, and 4.46 kW was recorded at 1200 rpm for all tested fuels: Diesel, 1LDiesel + 1ML DXP, 2LDiesel + 1ML DXP, and 3LDiesel + 1ML DXP,

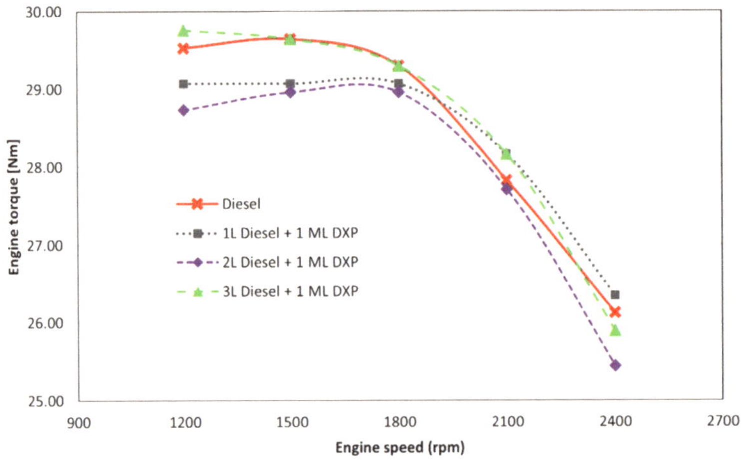

Figure 2.
Variation of BT according to engine speed at open throttle conditions for diesel and DXP blends.

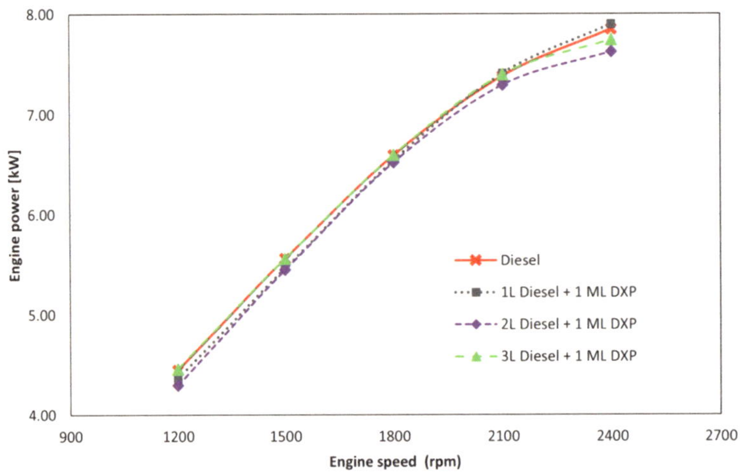

Figure 3.
Variation of BP according to engine speed at open throttle conditions for diesel and DXP blends.

respectively. The average BP values of Diesel, 1LDiesel + 1ML DXP, 2LDiesel + 1ML DXP, and 3LDiesel + 1ML DXP for all engine speed ranges (1200 to 2400 rpm) were 6.36, 6.33, 6.23, and 6.35 kW, respectively. The average BP decreased by 0.45, 2.02, and 0.23% for 1LDiesel + 1ML DXP, 2LDiesel + 1ML DXP, and 3LDiesel + 1ML DXP, respectively, compared to diesel.

3.3 Brake-specific fuel consumption (BSFC)

Figure 4 illustrates the variation in BSFC for all fuel blends operated under open throttle conditions with variable engine speed. The comparison between diesel and DXP blended fuel is shown in **Figure 4**. Generally, the BSFC values for all fuel blends first decline as engine speed increases, then slightly rise for higher engine rpms. More fuel was consumed initially to overcome inertial effects and keep the engine running.

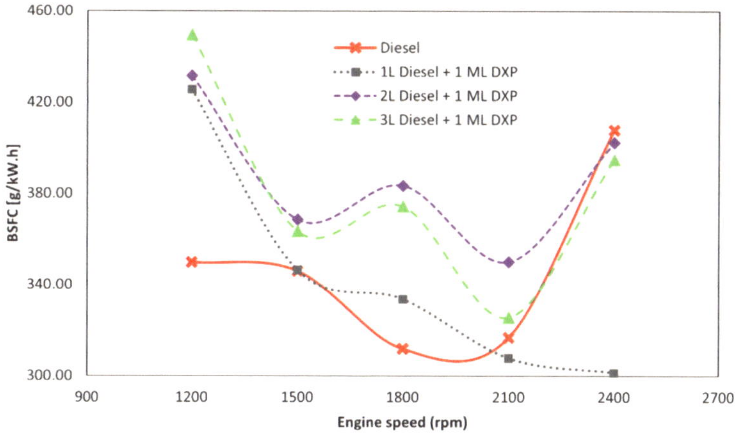

Figure 4.
Variation of BSFC according to engine speed at open throttle conditions for diesel and DXP blends.

The heat loss from the cylinder walls of the engine was more significant at lower engine speeds, which prompted higher fuel consumption to compensate for such losses. The minimum value of BSFC was found 311.89 kg/kWh at 1200 rpm for Diesel, 301.34 kg/kWh at 2400 rpm for 1LDiesel + 1ML DXP, 349.91 kg/kWh at 2100 rpm for 2LDiesel + 1ML DXP, 325.62 kg/kWh at 2100 rpm for 3LDiesel + 1ML DXP. The maximum value of BSFC was found 407.80 kg/kWh at 2400 rpm for diesel, 425.53 kg/kWh at 1200 rpm for 1LDiesel + 1ML DXP, 431.53 kg/kWh at 1200 rpm for 2LDiesel + 1ML DXP, 449.45 kg/kWh at 1200 rpm for 3LDiesel + 1ML DXP. The average BSFC values of diesel, 1LDiesel + 1ML DXP, 2LDiesel + 1ML DXP, and 3LDiesel + 1ML DXP for all engine speed ranges (1200 to 2400 rpm) were 346.46, 342.97, 387.09, and 381.53 kg/kWh, respectively. On average, it was found that the BSFC for 2LDiesel + 1ML DXP and 3LDiesel + 1ML DXP increased by 11.73 and 10.12% compared to diesel. However, the average BSFC for 1LDiesel + 1ML DXP decreased by 1.01% compared to diesel. The higher BSFC is mainly because of lower heating values and higher viscosity, while lower BSFC is due to better fuel atomization and improved air-to-fuel mixture. For higher volumetric efficiency, the required work during compression stoke decreases, which ultimately reflects through improved BSFC. Higher volatility and cetane index are the potential factors for a better air-fuel mixture, leading to a higher combustion efficiency. Better atomization, a higher calorific value, a shorter ignition delay, shorter burn periods, a high evaporation rate, and a larger reactive surface area with air can all enhance the combustion process. Because less fuel was burned during the premixed phase due to the decreased ignition delay, the BSFC decreased.

3.4 Brake thermal efficiency (BTE)

Figure 5 shows the BTE variance for all test fuels at open throttle with varying engine speeds (1200–2400 rpm). Due to the poor air-fuel combination and poor spray characteristics at high engine rpm, BTE for all test fuels increased until a specific speed limit and then decreased at higher engine speeds. The maximum value of BTE was found 34.47% at 1800 rpm for Diesel, 35.68% at 2400 rpm for 1 L Diesel +1ML DXP, 30.73% at 2100 rpm for 2 L Diesel +1ML DXP, 33.02% at 2100 rpm for 3 L Diesel +1ML DXP. The minimum value of BTE was found 26.37% at 2400 rpm for Diesel, 25.27% at 1200 rpm for 1 L Diesel +1ML DXP, 24.92% at 1200 rpm for 2 L Diesel

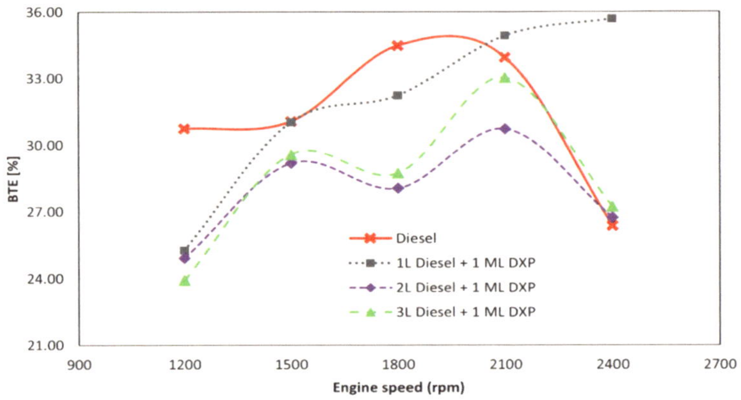

Figure 5.
Variation of BTE according to engine speed at open throttle conditions for diesel and DXP blends.

+1ML DXP, 23.92% at 1200 rpm for 3 L Diesel+1ML DXP. The average BTE values of diesel, 1 L Diesel +1ML DXP, 2 L Diesel +1ML DXP, and 3 L Diesel +1ML DXP for all engine speed ranges (1200 to 2400 rpm) were 31.22, 31.83, 27.92.66, and 28.50%, respectively. On average, it was found that the BTE for 2 L Diesel +1ML DXP and 3 L Diesel +1ML DXP decreased by 10.85 and 9.01% compared to diesel. However, the average BTE for 1 L Diesel +1ML DXP increased by 1.62% compared to diesel. The high viscosity results in poor atomization and results in lower BP and BTE. The longer ignition delay time resulted in higher BTE due to an improved combustion process and shorter combustion duration. Oxidizing agents increase thermal efficiency for oxygenated ternary fuel blends by introducing oxygen into a fuel-rich zone, which improves combustion in the engine cylinder. The full combustion process and high oxygen content, which will enhance the fuel-air ratio in the fuel-rich area during diffusive combustion, maybe the cause of the BTE enhancement. High in-cylinder pressure, high cetane number, high heat release rate (HRR), and the shortest ignition delay all contribute to the greater BTE.

3.5 CO emissions

Figure 6 shows how CO emissions affect engine speed for tested fuels. For every fuel that was tested, the highest CO emissions were measured at 1200 rpm, and the lowest CO emissions were measured at 2400 rpm. Fuel's partial combustion and partial oxidation of carbon atoms result in CO emissions. CO emissions can be varied by changing the air-fuel ratio in the engine cylinder. CO emissions show an inverse relation with engine speeds. The transformation rate of CO to CO_2 primarily rises due to lower air-fuel ratio and elevated temperature at higher engine speeds, which ultimately results in lower CO emissions, as portrayed in **Figure 6**. Due to complete combustion during the combustion process in diesel engines, all evaluated fuel samples showed a considerable reduction in CO emissions compared to commercial diesel. The average CO emissions for diesel, 1LDiesel + 1ML DXP, 2LDiesel + 1ML DXP, and 3LDiesel + 1ML DXP between (1200–2400 rpm) were 3.58, 3.41, 3.51, and 3.24 vol.%, respectively. On average, all DXP blends portrayed a significant decline in CO emissions of 4.81, 2.21, and 5.12% for 1LDiesel + 1ML DXP, 2LDiesel + 1ML DXP, and 3LDiesel + 1ML DXP, respectively, in comparison with diesel fuel. Among all

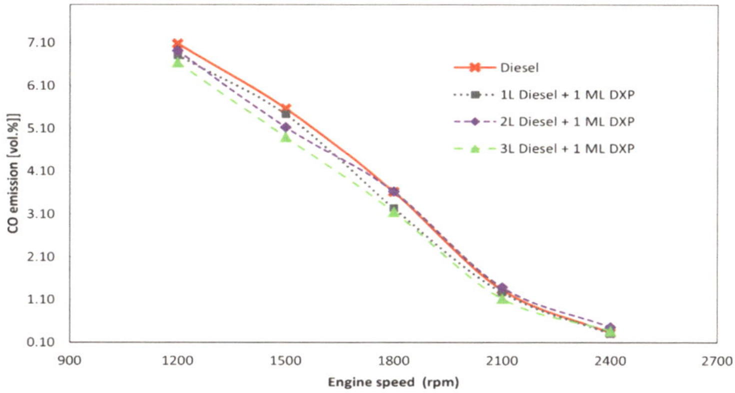

Figure 6.
Variation of CO emissions according to engine speed at open throttle conditions for diesel and DXP blends.

the evaluated fuels, the DXP blends portrayed a significant reduction due to the high oxygen content in their molecular structure.

Additionally, reduced density and viscosity improved fuel atomization. Lower CO emissions resulted from the fuel burning completely, thanks to the increased oxygen content. DXP fuel blends' higher oxygen concentration guarantees CO oxidation in fuel-rich regions within the cylinder, lowering CO emissions.

3.6 CO_2 emissions

Figure 7 displays the variation in CO_2 emissions for all test fuels concerning engine speeds. The general increasing trend can be observed in CO_2 emissions as engine speed increases. It indicates improved combustion at higher engine speeds. The average CO_2 emissions for diesel, 1LDiesel + 1ML DXP, 2LDiesel + 1ML DXP, and 3LDiesel + 1ML DXP between (1200–2400 rpm) were 9.19, 9.24, 8.93, and 8.98 vol.%, respectively. On average, the 2LDiesel + 1ML DXP and 3LDiesel + 1ML DXP portrayed a significant reduction in CO_2 emissions of 2.81 and 2.28%, respectively, compared with diesel fuel. However, the CO_2 emissions for 1LDiesel + 1ML DXP increased by

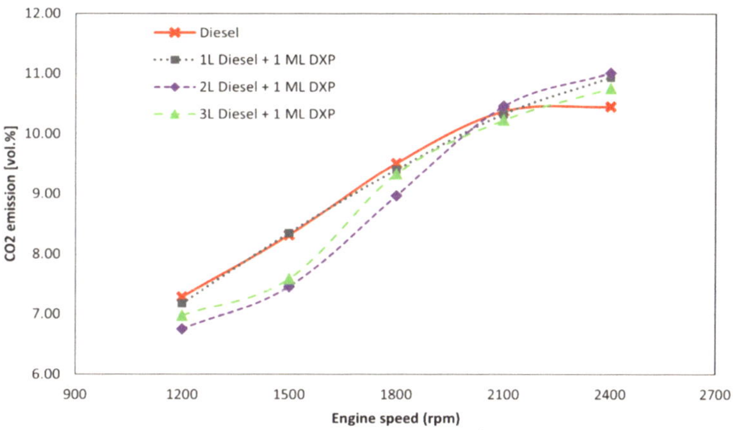

Figure 7.
Variation of CO_2 emissions according to engine speed at open throttle conditions for diesel and DXP blends.

0.53% compared to diesel. A possible reason for lower CO_2 emissions is improved combustion, which results in better fuel burning. The blends formulated to have better vaporization and combustion properties ensure more complete fuel burning, which reduces CO_2 emissions. This can include fuels with higher cetane numbers or those specifically designed for cleaner combustion.

3.7 HC emissions

Figure 8 shows how HC emissions vary as engine speed increases. Fuel atomization, engine operating conditions, and fuel characteristics are the primary determinants of HC emissions. On average, the HC emissions were 32.84, 29.68, 32.40, and 28.24 ppm for diesel, 1LDiesel + 1ML DXP, 2LDiesel + 1ML DXP, and 3LDiesel + 1ML DXP between 1200 and 2400 rpm were recorded, respectively. All DXP blends significantly reduced HC emissions compared to commercial diesel. On average, the DXP blends demonstrated significant decrements in HC emissions of 9.62, 1.34, and 14.01% for 1LDiesel + 1ML DXP, 2LDiesel + 1ML DXP, and 3LDiesel + 1ML DXP, respectively, compared to commercial diesel. The improved burning of blended fuels is responsible for this notable decrease in HC emissions. The oxidation process of unburned hydrocarbons was caused by the high in-cylinder temperature and high heat release rate, which reduced the HC emissions.

3.8 NO emissions

Figure 9 shows the NO emissions for each tested fuel. For all gasoline blends, the graph showed an increasing tendency as speed increases. The average NO emissions of 359.20, 354.28, 349.36, and 356.32 ppm were obtained for diesel, 1LDiesel + 1ML DXP, 2LDiesel + 1ML DXP, and 3LDiesel + 1ML DXP, respectively. The temperature and pressure of in-cylinder combustion are the primary causes of NO production. Because of the elevated cetane number and oxygen concentration, a high in-cylinder temperature causes a high NOx generation. Compared to diesel fuel, 1LDiesel + 1ML DXP, 2LDiesel + 1ML DXP, and 3LDiesel + 1ML DXP showed an average NO emission reduction of 1.37, 2.74, and 0.80%, respectively.

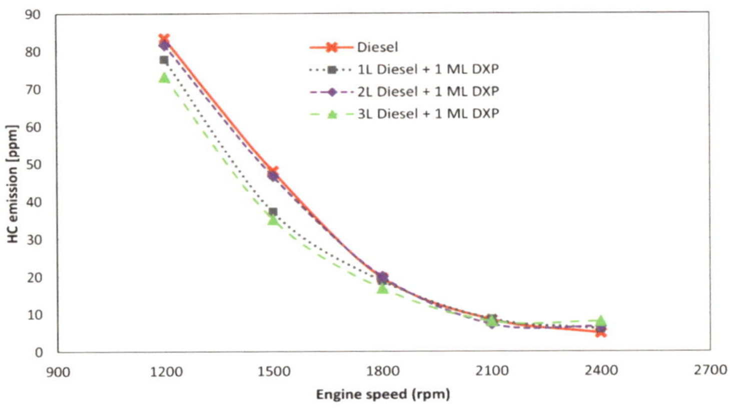

Figure 8.
Variation of HC emissions according to engine speed at open throttle conditions for diesel and DXP blends.

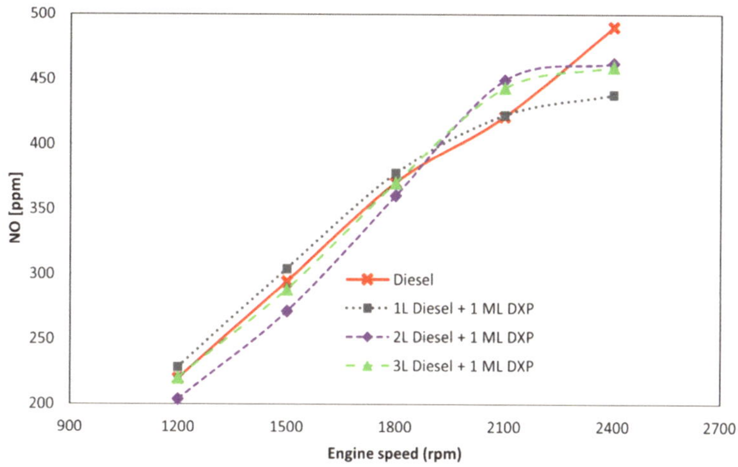

Figure 9.
Variation of NO emissions according to engine speed at open throttle conditions for diesel and DXP blends.

4. Conclusions

DXP-Diesel fuel blends produced 0.48 to 1.83% less torque than diesel fuel except 3LDiesel + 1ML DXP fuel blends that produce 0.24% higher torque than diesel fuel. However, DXP fuel blends, on average, produced 0.23 to 2.02% lower brake power than diesel fuel.

On average, it was found that the BSFC for 2LDiesel + 1ML DXP and 3LDiesel + 1ML DXP increased by 11.73 and 10.12% compared to diesel. However, the average BSFC for 1LDiesel + 1ML DXP decreased by 1.01% compared to diesel. BTE for 2 L Diesel +1ML DXP and 3 L Diesel +1ML DXP decreased by 10.85 and 9.01% compared to diesel. However, the average BTE for 1 L Diesel +1ML DXP increased by 1.62% compared to diesel.

All DXP blends portrayed a significant decline in CO emissions of 4.81, 2.21, and 5.12% for 1LDiesel + 1ML DXP, 2LDiesel + 1ML DXP, and 3LDiesel + 1ML DXP, respectively, compared to diesel fuel. The 1LDiesel + 1ML DXP showed a reduction in CO_2 emissions of 0.44% compared to Diesel. However, the CO_2 emissions for 3LDiesel + 1ML DXP increased by 0.26% compared to Diesel and no change in CO_2 emissions was observed for 2LDiesel + 1ML DXP.

The DXP blends portrayed significant reduction in HC emissions of 9.62, 1.34, and 14.01% for 1LDiesel + 1ML DXP, 2LDiesel + 1ML DXP, and 3LDiesel + 1ML DXP, respectively, compared to commercial diesel.

Compared to diesel fuel, 1LDiesel + 1ML DXP, 2LDiesel + 1ML DXP, and 3LDiesel + 1ML DXP showed an average NO emission reduction of 1.37, 2.74 and 0.80%, respectively.

Acknowledgements

I want to express my gratitude to UTS and FuelnGo Pty Ltd. for their support in making this project possible and successful. A special thanks to Prof. Long D. Nghiem, Director of CTWW, UTS; Petra Beaker, Business Research Engagement

Associate, UTS, for their guidance and encouragement in the development of this project with DXP Global Australia (FuelnGo Pty Ltd). I would also like to thank RMIT, UNSW, UM (Malaysia), and other laboratories in Sydney for helping with the experimental work for this project.

Conflict of interest

The authors declare no conflict of interest.

Appendices and nomenclature

BT	Brake torque
BTE	Brake thermal efficiency
BSFC	Brake-specific fuel consumption
CO	carbon monoxide
CO_2	carbon dioxide
DPF	diesel particulate filter
Deg	degree
HC	hydrocarbon
kw	Kilo-watt
Kg/cm^2	kilogram per centimeter square
L	litre
mm	millimetre
NO	oxide of nitrogen
PM	particulate matter
ppm	parts per million
SAE	society of automotive engineer

Author details

Md Abul Kalam[1*], T.M. Indra Mahlia[1], Arridina Susan Silitonga[1], Islam Md Rizwanul Fattah[1], Md Mofijur Rahman[1], Mohammad Rashed Mia[1], Muhammad Ali Ijaz Malik[1], Nicholas Moulis[2], Anthony Witts[3], James Witts[3] and Jonathan Sherlock[3]

1 University of Technology Sydney, Australia

2 FuelnGo Pty Ltd, Australia

3 Diesel X-tra Pty Ltd Trading as DXP Global, Australia

*Address all correspondence to: mdabul.kalam@uts.edu.au

IntechOpen

© 2025 The Author(s). Licensee IntechOpen. This chapter is distributed under the terms of the Creative Commons Attribution License (http://creativecommons.org/licenses/by/4.0), which permits unrestricted use, distribution, and reproduction in any medium, provided the original work is properly cited.

References

[1] Wakil MA, Kalam MA, Masjuki HH, Atabani AE, Rizwanul Fattah IM. Influence of biodiesel blending on physicochemical properties and importance of mathematical model for predicting the properties of biodiesel blend. Energy Conversion and Management. 2015;**94**:51-67 [Online]. Available from: http://www.sciencedirect.com/science/article/pii/S0196890415000473

[2] Cecrle E et al. Investigation of the effects of biodiesel feedstock on the performance and emissions of a single-cylinder diesel engine. Energy & Fuels. 2012;**26**(4):2331-2341

[3] Absi Halabi M, Al-Qattan A, Al-Otaibi A. Application of solar energy in the oil industry—Current status and future prospects. Renewable and Sustainable Energy Reviews. 2015;**43**:296-314 [Online]. Available from: http://www.sciencedirect.com/science/article/pii/S1364032114009617

[4] Aransiola E, Ojumu T, Oyekola O, Madzimbamuto T, Ikhu-Omoregbe D. A review of current technology for biodiesel production: State of the art. Biomass and Bioenergy. 2014;**61**:276-297

[5] Vardoulakis S et al. Impact of climate change on the domestic indoor environment and associated health risks in the UK. Environment International. 2015;**85**:299-313

[6] REN21. Renewables 2019 Global Status Report. Vol. 2019. Paris: REN21; 2019. 978-3-9818911-3-3. [Online]. Available from: https://www.ren21.net/gsr-2019/chapters/chapter_01/chapter_01/#sub_1_1

[7] REN21. Renewables 2020 Global Status Report. Paris: REN21; 2020. 978-3-948393-00-7. [Online]. Available from: https://www.ren21.net/gsr-2020/chapters/chapter_01/chapter_01/#sub_4 [Accessed: February 2, 2021]

[8] REN21. Renewables 2018 Global Status Report. Paris: REN21; 2018. 978-3-9818911-3-3. [Online]. Available from: https://www.ren21.net/gsr-2018/chapters/chapter_01/chapter_01/

[9] Uusitalo V, Leino M, Kasurinen H, Linnanen L. Transportation biofuel efficiencies from cultivated feedstock in the boreal climate zone: Case Finland. Biomass and Bioenergy. 2017;**99**:79-89

[10] Gavhane RS et al. Effect of zinc oxide nano-additives and soybean biodiesel at varying loads and compression ratios on VCR diesel engine characteristics. Symmetry. 2020;**12**(6):1042

Chapter 5

Driving Characteristics of Automobiles When Overtaking Bicycles in Tunnels

Shuichi Matsumoto

Abstract

Japan has a high fatality rate on road sections. This trend has been observed abroad. Several studies have focused on instrumental bicycles to analyze phenomena related to safety on bicycles since 2013, particularly in Europe and the United States. Research on vehicle passing distances to bicycles began in 1977 and has been followed several studies. It has been reported that the dark, narrow, and monotonous spaces in tunnels make drivers. As there remain several unanswered questions regarding the awareness of bicyclists in tunnels and the passing distance between bicycles and vehicles, this study conducted comparative experiments inside and outside of tunnels using the instrumental bicycle. The following findings were obtained from this study. 1. The passing distance tended to be wider outside tunnels than inside tunnels. 2. Bicycles inside tunnels tend to have shorter passing distances than those outside tunnels. 3. Bicycles tend to have shorter passing distances than vehicles.

Keywords: passing distance, overtaking cyclists, instrument bicycle, vehicle-bicycle interaction, tunnel

1. Introduction

Japan has a high fatality rate on road sections. This trend has been observed abroad as well, with the Transport for London [1] reporting that crashes involving passing motor vehicles were the leading cause of death in bicycle accidents. A study by the National Highway Traffic Safety Administration (NHTSA) of the United States Department of Transportation [2] reported that 57% of all bicycle fatalities and injuries in 2013 occurred outside of intersections. In Sweden, cyclists were found to be the most vulnerable road users, and cyclists were at high risk of serious injury in collisions between bicycles and motor vehicles [3]. Thus, in traffic accidents, bicycle users are among the most vulnerable in many countries, leading to serious accidents.

Several studies have focused on instrumental bicycles to analyze phenomena related to safety on bicycles since 2013, particularly in Europe and the United States [4]. Research on vehicle passing distances to bicycles began with Kroll and Ramey's study [5] in 1977 and has been followed several studies.

Many studies on passing distances have explored the factors that contribute to passing distances based on road structure, user attributes, and other factors. However, there exist several factors that yet to be appropriately examined. A study by Shackel and Parkin [6] stated that passing distances tended to be wider on multi-lane roads than on single-lane roads. However, a study by Apasnore et al. [7] found no statistically significant difference. Regarding the effect of the presence of on-street parking on the passing distance, the study by Apasnore et al. [7] concluded that the passing distance increased, whereas Beck et al. [8] found that the passing distance decreased. In a study by Walker [9], drivers demonstrated that the appearance of the bicyclist and other factors significantly affected the passing distance, with one example showing that the passing distance was greater when the bicyclist was female. In contrast to this result, the study by Haworth et al. [10] found no statistically significant difference. The experiment by Evans et al. [11] reported a 7.6 cm shortening.

Caliendo [12] reported that while several studies have focused on traffic accidents on roads, those on accidents in tunnels are scarce. In case of studies on passing distances, research cases considering tunnels (**Table 1**) are scarce. Regarding bicycle accidents on road sections, a project wherein the author participated themselves, compared the serious accident rate of bicycle accidents on road sections based on

Country	Year	Number of roads measured	Passing events recorded	Research Scope	Distance measurement technology	Reference	Average passing distance
UK	2010	6	671	Out of Scope	Bike mounted video footage	Parkin and Meyers [13]	1.1–1.6 m
Canada	2015	4	5227	unknown	Ultrasonic rack mounted	Mehta et al. [14]	1.3–2.8 m
NewZealand	2016	6	6268	Out of Scope	Lidar sensor	Balanovic et al. [15]	2.05 m
USA	2012	37	568	unknown	Bike mounted video footage	Love et al. [16]	1.45 m
UK	2014	1	5690	Out of Scope	Ultrasonic mounted low on bike	Walker et al. [17]	—
Taiwan	2013	1	1380	Out of Scope	Ultrasonic mounted low on the bike	Chuang et al. [18]	1.68 m
USA	2012	1	1151	unknown	Ultrasonic rack mounted	Chapman and Noyce [19]	1.95 m
Sweden	2016	1	145	Out of Scope	Lidar sensor	Dozza et al. [3]	1.8–3 m
Spain	2017	7	2928	Out of Scope	Laser sensor	Llorca et al. [20]	1.5 m
Australia	2019	unknown	18,527	Out of Scope	Ultrasonic Saddle mount	Beck et al. [8]	1.73 m
Australia	2021	unknown	16,476	unknown	ultrasonic distance sensor 20 Hz	Mackenzie et al. [21]	—
CANADA	2020	unknown	3591	unknown	Ultrasonic handle mount	Henao and Apparicio [22]	1.76 m

Country	Year	Number of roads measured	Passing events recorded	Research Scope	Distance measurement technology	Reference	Average passing distance
USA	2021	33	2838	Out of Scope	Ultrasonic handle mount	Feizi et al. [23]	1.81 m
Australia	2018	10	1846	unknown	video camera attached to roadside poles or sign posts	Kumar et al. [24]	1.86 m
UK	2014	unknown	463	Out of Scope	Ultrasonic rack mounted	Shackel and Parkin [6]	1.70 m
USA	2020	3	120	Out of Scope	Driving Simulator	Herrera et al. [25]	1.85 m
Italy	2021	1	134	Out of Scope	Driving Simulator	Rossi et al. [26]	1.41–1.42m
JAPAN	2018	1	288	Out of Scope	Driving Simulator	Francesco et al. [27]	1.27–1.46m
FRANCE	2020	6	1296	Out of Scope	Driving Simulator	Mecheri et al. [28]	0.8–1.54 m
Australia	2018	15	1113	Out of Scope	video camera attached to roadside poles or sign posts	Haworth et al. [10]	1.41–1.63m

Table 1.
Summary of research about passing distance.

	Slight injury	Fatal injury	Death	Total	Probability of serious accidents
Tunnel	212	85	8	305	30.5%
Bridge	191	47	10	248	23.0%
Curve	2722	453	30	3205	15.1%
Other	58,611	4956	295	63,862	8.2%

Table 2.
Probability of serious accidents by road section.

Japanese accident data. They found that the serious accident rate was higher in tunnel sections (**Table 2**). Further, it has been reported that the dark, narrow, and monotonous spaces in tunnels make drivers feel uneasy [29]. As there remain several unanswered questions regarding the awareness of bicyclists in tunnels and the passing distance between bicycles and vehicles, this study conducted comparative experiments inside and outside of tunnels.

2. Method

2.1 Overview of instrumental bicycle

In this study, the instrumental bicycle shown in **Figure 1** was used. The speed of the vehicle was measured using a camera mounted on the bicycle. Measurements were

Figure 1.
Outline of instrumental bicycle.

performed 3 m behind the rear camera while passing through a distance of 0.91 m from the rear camera to the front camera and ended 3 m ahead of the front camera (**Figure 2**). The time to pass through the distance was measured and calculated based on the time difference using Eq. 1.

$$v_{car} = \frac{6.91}{t_{front} - t_{rear}} \times 3.6 + v_{bike} \quad (1)$$

where vcar (km/h) is the speed of the vehicle, vbike (km/h) is the speed of the bicycle, tfront is the measurement start time of the front camera, and trear is the measurement end time of the rear camera. To measure the passing distance, an ultrasonic sensor (HC-SR04) installed horizontally on the right side of the rear basket of the bicycle was used to continuously record the distance from the bicycle to the side of the vehicle at approximately 7 Hz, and the shortest value among multiple values measured during overtaking was used.

2.2 Experimental site subjects

In this study, driving experiments were conducted on the Yoshiura Tunnel—Shin Iizumi Tunnel section of National Route 16 in Yokosuka City. (**Figure 3**) This road has two lanes on each side with a speed limit of 40 km/h and a travel distance of 3.2 km. There are nine tunnels in this section. A summary of each tunnel is presented in **Table 3**. Two types of lighting were present in the experimental tunnels: tunnels with LEDs and sodium lamps (hereafter referred to as "mixed") and tunnels with sodium lamps (hereafter referred to as "yellow"). The traffic volume was 20,609 vehicles during the 12-h daytime period (8:00 AM to 8:00 PM). For the road section, data for six locations are shown in **Table 3**.

Since the same road was used for the inside and outside of the tunnel, no significant deviations in traffic volume and travel speeds were observed.

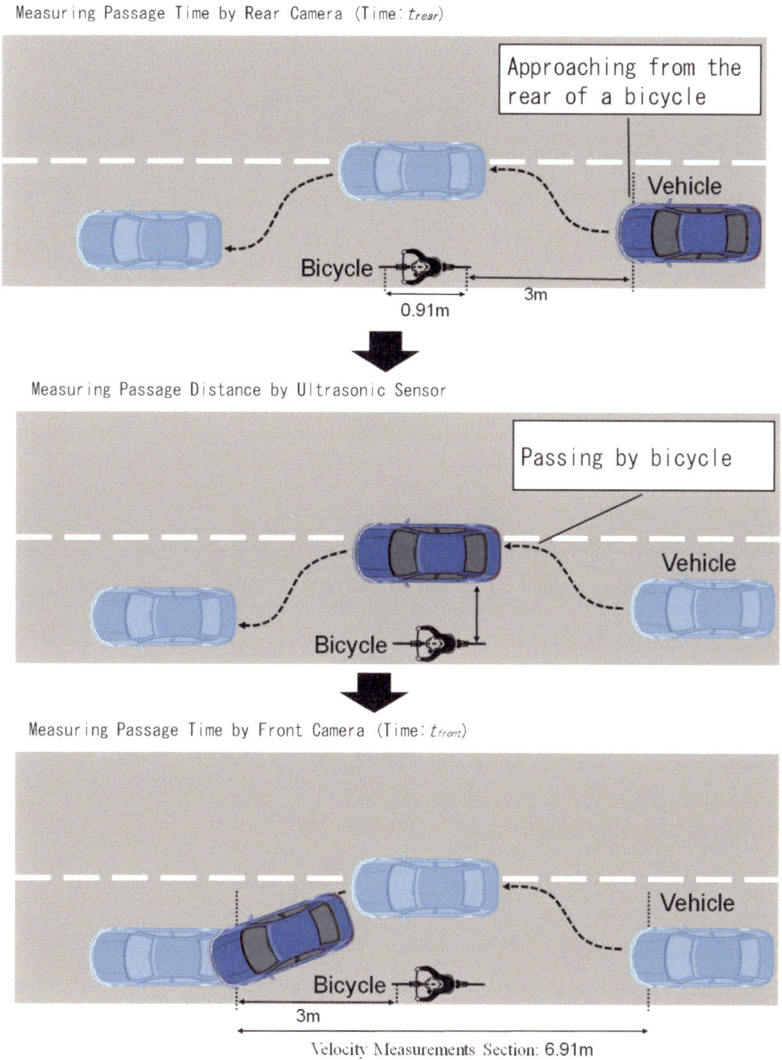

Figure 2.
Outline of measuring vehicle speed.

The location of this experiment was selected because of the existence of several tunnels in succession. It is also close to the author's workplace. Therefore, it was selected as the site for the experiment because it was thought that data could be collected efficiently. However, these results were obtained in a tunnel with two lanes on each side, and it is necessary to conduct additional tests in various tunnels in the future to confirm whether this is a general characteristic.

This study involved 11 persons (8 males and 3 females) between the ages of 19–22. Prior to the start of the experiment, the subjects were fully briefed on the disadvantages to the participants, privacy considerations, and freedom not to participate in the experiment. The duration of the experiment was 2 h per subject.

Subjects were given sufficient explanations regarding (1) disadvantages to the subjects caused by the experiment, (2) privacy considerations, and (3) freedom not to

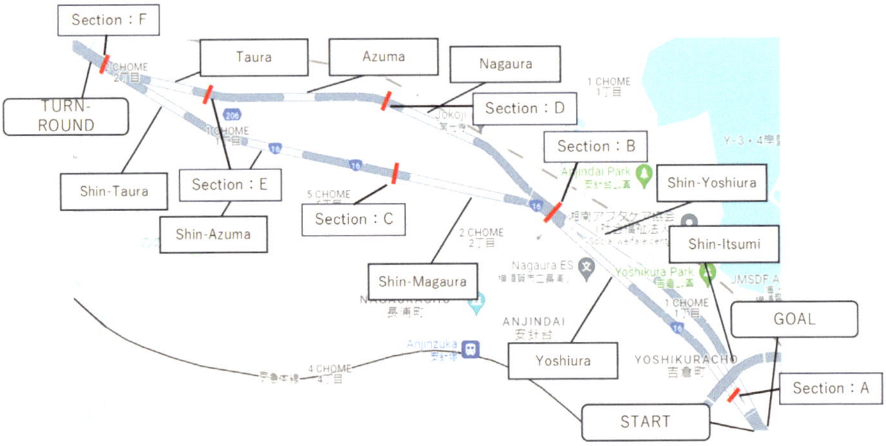

Figure 3.
Overview of experiment location.

Section ID	Number of lane	Width of road (m)	Speed zone (km/h)	Structure
A	2	7.0	40	One-way
B	2/2	8.4/7.0	40	Two-way
C	2	8.3	40	One-way
D	2	9.9	40	One-way
E	2	7.5	40	One-way
F	2/2	7.5/7.5	40	Two-way

Table 3.
Road section data about outside the tunnel.

participate in the experiment, and their consent to participate in the experiment was obtained in the informed consent session before the experiment began. This experiment was approved by the Research Ethics Review Committee of Bunkyo University (Reception No. 2022-02). In addition, the experiment was conducted in accordance with the Bunkyo University Research Ethics Regulations and the Bunkyo University Regulations on the Protection of Personal Information.

In this experiment, subjects rode a probe bicycle and were instructed to ride on the outside line of the roadway as much as possible. In order to collect the sense of danger felt by the bicyclists when a car or motorcycle overtook their bicycles, a piece of paper indicating four levels of danger (0: safe, 1: somewhat dangerous, 2: dangerous, and 3: very dangerous) was attached to the front basket of the bicycle. When the subject felt "somewhat dangerous" or more during overtaking, he or she was asked to vocalize the number of the sense of danger.

3. Result

This experiment collected 471 samples of overtaking maneuvers. A breakdown of these and the passing distances and bicycle and vehicle speeds are summarized in **Table 4**.

Characteristic		N	Passing distance (cm)		Velocity of motor vehicle (km/h)		Velocity of bicycle (km/h)	
			Average	SD	Average	SD	Average	SD
Sex of Cyclist	Male	311	115.4	33.28	44.7	10.09	17.1	3.28
	Female	160	112.3	29.38	44.3	9.10	15.4	2.85
Inside/ Outside of Tunnels	Inside of Tunnels (Yellow)	96	106.1	24.52	44.5	10.26	15.7	3.53
	Inside of Tunnels (Mix)	131	102.0	28.92	44.1	8.38	16.2	2.79
	Outside of Tunnel	241	124.5	33.11	44.8	10.30	17.0	3.29
Motor Vehicle Type	Light Car	232	111.6	28.90	44.2	10.14	16.2	3.09
	Sedan	157	110.4	33.38	43.6	7.64	16.9	3.54
	Truck and Bus	22	103.3	21.05	41.9	12.80	16.1	2.47
	Motorbike	60	139.6	32.00	49.5	10.56	16.7	3.21
Relation of Oncoming Vehicle	over 5 sec	46	119.6	34.89	47.7	8.52	16.8	3.03
	under 5 sec	425	113.8	31.68	44.2	9.83	16.5	3.27

Table 4.
Summary of result.

3.1 Passing distance

The passing distances inside and outside the tunnel are presented in **Figure 4**. One-way analysis of variance for the passing distance exhibited a significant difference in the main effect of road structure ($F(2,469) = 4.43$, $p < 0.001$). Multiple comparisons using the Tukey method indicated a significant trend between "mixed" (LED and sodium lights) and "outside" and "yellow" (sodium lights) and " outside " was shown. These results indicate that there was no effect of lighting inside the tunnel on the passing distance; however, the passing distance tended to be wider inside the tunnel compared to that outside the tunnel.

3.2 Risk perception

Figure 5 shows the percentage of danger for each lighting in the tunnel. As evident, the percentage of mixed "very dangerous" accounted for more than that of yellow, outside the tunnel. Further, that the non-safety rating was 50.8% of the total, or approximately half of the total. However, the percentage of safety outside the tunnel was found to be no different than that inside the tunnel. Next, we plotted the kernel density estimates of the overtaking passing distances at which subjects felt unsafe inside and outside the tunnel. The results indicated that the passing distance in the tunnel for overtaking that the bicyclists found dangerous was shorter (**Figure 6**).

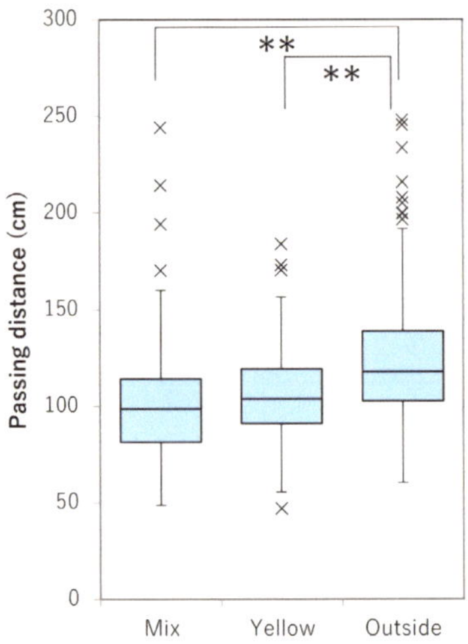

Figure 4.
Passing distance by road type.

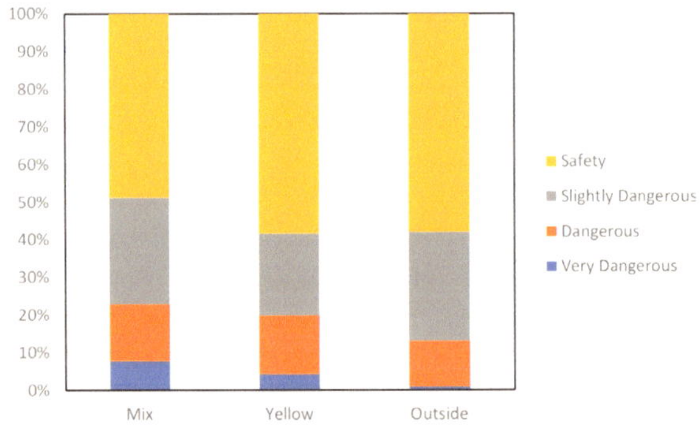

Figure 5.
Percentage of danger for each road.

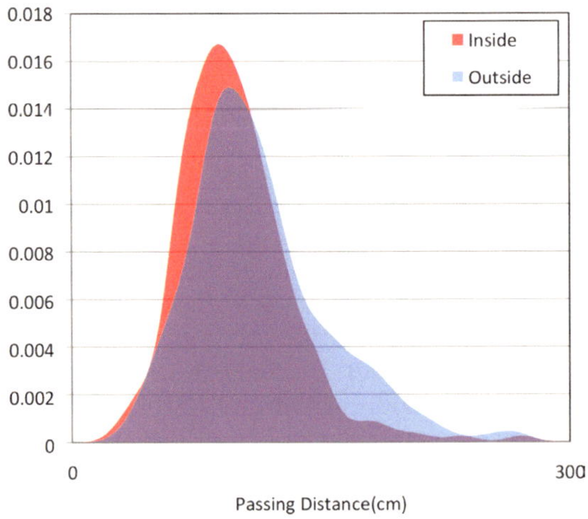

Figure 6.
Distribution of passing distances for overtaking that felt dangerous.

4. Conclusion

In Japan, the percentage of serious accidents involving bicycles in tunnel sections is higher than in road sections. However, even from an international perspective, research regarding passing distances for bicycles in tunnels is limited. This study measured the passing distance when a vehicle overtook a bicycle using an instrumental bicycle on a two-lane arterial road with a series of tunnels. Consequently, data on the risk perception by bicycle users were collected.

The following findings were obtained from this study.

1. The passing distance tended to be wider outside tunnels than inside tunnels.

2. Bicycles inside tunnels tend to have shorter passing distances than those outside tunnels.

3. Bicycles tend to have shorter passing distances than vehicles.

In this study, we compared passing distances on a two-lane road in one direction, inside and outside of tunnels. Studies of passing distances have included the effects of various roadway factors on passing distances, such as Stülpnagel et al. [29] and Shackel et al. [15]. The results of this study show a trend on a two-lane road, and it is necessary to verify whether similar results can be obtained in many tunnels in the future. In overtaking studies, driving simulators can be used to freely create road and traffic conditions, allowing experimenters to conduct comparative experiments (Goddard et al. [30], Farah et al. [31], Piccinini et al. [27]) focused on the environment we wish to verify. Furthermore, it is also important to examine the mechanism of overtaking a bicycle by a vehicle, taking into account the effects of driver's risk awareness (Rasch et al. [32]) and cyclist's stress. Based on these verifications, we plan to increase our knowledge that will lead to the development of a comfortable driving space for vehicles in tunnels.

Acknowledgements

This research was conducted as part of "Research on cycle route sequencing using bicycle DX", the commissioned research of "National Institute for Land and Infrastructure Management" under technology research and development system of "The Committee on Advanced Road Technology" established by MLIT, Japan. The author would like to thank Dr. Nagahiro Yoshida from Osaka Metropolitan University, Mr. Keidai Kobayashi from KI-STAR REAL ESTATE CO., LTD. This research was carried out as "Research on cycle route sequencing using bicycle DX", the commissioned research of "National Institute for Land and Infrastructure Management" under technology research and development system of "The Committee on Advanced Road Technology" established by MLIT, Japan.

Conflict of interest

The author declares no conflict of interest.

Author details

Shuichi Matsumoto
Bunkyo University, Chigasaki, Japan

*Address all correspondence to: shuichi@bunkyo.ac.jp

IntechOpen

© 2025 The Author(s). Licensee IntechOpen. This chapter is distributed under the terms of the Creative Commons Attribution License (http://creativecommons.org/licenses/by/4.0), which permits unrestricted use, distribution, and reproduction in any medium, provided the original work is properly cited.

References

[1] Transport for London Street Management. Pedal Cyclist Casualties in Greater London. London: Transport for London; 2005. p. 2

[2] NHTSA's National Center for Statistics and Analysis. Washington, D.C: Traffic safety facts 2013; 2015. p. 10

[3] Dozza M, Schindler R, Piccinini GB, Karlsson J. How do drivers overtake cyclists? Accident Analysis and Prevention. 2016;**88**:29-36

[4] Gadsby A, Watkins K. Instrumented bikes and their use in studies on transportation behaviour, safety, and maintenance. Transport Reviews. 2020;**40**(6):774-795

[5] Kroll B, Ramey M. Effects of bike lines on driver and bicyclist behavior. Transportation Engineering Journal of ASCE. 1977;**103**:243-256

[6] Shackel SC, Parkin J. Influence of road markings, lane widths and driver behaviour on prox-imity and speed of vehicles overtaking cyclists. Accident Analysis and Prevention. 2014;**73**:100-108

[7] Apasnore P, Ismail K, Kassim A. Bicycle-vehicle interactions at mid-sections of mixed traffic streets: Examining passing distance and bicycle comfort perception. Accident Analysis & Prevention. 2017;**106**:141-148

[8] Beck B, Chong D, Olivier J, Perkins M, Tsay A, Rushford A, et al. How much space to drivers provide when passing cyclists? Understanding the impact of motor vehicle and infrastructure characteristics on passing distance. Accident Analysis and Prevention. 2019;**128**:253-260

[9] Walker I. Drivers overtaking bicyclists: Objective data on the effects of riding position, helmet use, vehicle type and ap-parent gender. Accident Analysis and Prevention. 2007;**39**(2):417-786

[10] Haworth N, Heesch CK, Schramm A, Debnath KA. Do Australian drivers give female cyclists more room when passing? Journal of Transport & Health. 2018;**9**:203-211

[11] Evans I, Pansch J, Singer-Berk L, Lindsey GP. Factors affecting vehicle passing distance and encroachments while overtaking cyclists. Institute of Transportation Engineers. ITE Journal. 2018;**88**(5):40-45

[12] Caliendo C, Guglielmo M, Guida M. A crash-prediction model for road tunnels. Accident Analysis and Prevention. 2013;**55**:107-115

[13] Parkin J, Meyers C. The effect of cycle lanes on the proximity between motor traffic and cycle traffic. Accident; Analysis and Prevention. 2010;**42**:159-165

[14] Mehta K, Mehran B, Hellinga B. Analysis of lateral distance between motorized vehicles and cyclists during overtaking maneuvers. In: Proceedings of the Transportation Research Board 94th Annual Meeting. Washington DC; 2015. Available from: https://trid.trb.org/View/1336382

[15] Balanovic J, Davison A, Thomas J, Bowie C, Frith B, Lusby M, Burton J. Overtaking Gap Law for Motorists Overtaking Cyclists in New Zealand. Available from: https://www.nzta.govt.nz/assets/Walking-Cycling-and-Public-Transport/docs/

Minimum-Overtaking-Gap-Feasibility-Study-FINAL.pdf [Accessed: 25 January 2024]

[16] Love CD, Breaud A, Burns S, Margulies J, Romano M, Lawrence R. Is the three-foot bicycle passing law working in Baltimore, Maryland ? Accident Analysis and Prevention. 2012;**48**:451-456

[17] Walker I, Garrard I, Jowitt F. The influence of a bicycle commuter's appearance on drivers' overtaking proximities: An on-road test of bicyclist stereotypes, high-visibility clothing and safety aids in the United Kingdom. Accident Analysis and Prevention. 2014;**64**:69-77

[18] Chuang K, Hsu LC, Doong JM. The use of a quasi-naturalistic riding method to investigate bicyclists' behaviors when motorists pass. Accident Analysis and Prevention. 2013;**56**:32-41

[19] Chapman J, Noyce D. Observations of driver behavior during overtaking of bicycles on rural roads. Journal of the Transportation Research Board. 2012;**2321**:38-45

[20] Llorca C, Angel-Domenech A, Agustin-Gomez F, Garcia A. Motor vehicles overtaking cyclists on two-lane rural roads: Analysis on speed and lateral clearance. Safety Science. 2017;**92**:302-310

[21] Mackenzie JRR, Dutschke JK, Ponte G. An investigation of cyclist passing distances in the Australian Capital Territory. Accident Analysis and Prevention. 2021;**154**:1-10

[22] Henao A, Apparicio P. Dangerous overtaking of cyclists in Montréal. Safety. 2022;**8**(16)

[23] Feizi A, Oh I, Kwigizile J, Joo S. Cycling environment analysis by bicyclists' skill levels using instrumented probe bicycle. International Journal of Sustainable Transportation. 2020;**14**:722-732

[24] Debnath A, Haworth N, Schramm A, Kristiann C, Somoray H. Factors influencing noncompliance with bicycle passing distance laws. Accident Analysis and Prevention. 2018;**115**:137-142

[25] Herrera N, Scott A, Wolshon B. Driver compliance and safety effects of three-foot bicycle passing laws. Transportation Research Interdisciplinary Perspectives. 2020;**6**:1-8

[26] Rossi R, Orsini F, Tagliabue M, Stasi L, Gastaldi G. Evaluating the impact of real-time coaching programs on drivers overtaking cyclists. Transportation Research Part F. 2021;**78**:74-80

[27] Piccinini G, Moretto C, Zhou H, Itoh N. Influence of oncoming traffic on drivers' overtaking of cyclists. Transportation Research Part F. 2018;**59, Part A**:378-388

[28] Mecheria S, Roseyb F, Lobjois R. Manipulating constraints on driver-cyclist interactions in a fixed travel space: Effects of road configuration on drivers' overtaking behavior. Safety Science. 2020;**123**:1-11

[29] Stulpnagel R, Hologa R, Riach N. Cars overtaking cyclists on different urban road types -expectations about passing safety are not aligned with observed passing distances. Transportation Research Part F: Psychology and Behaviour. 2022;**89**:334-346

[30] Goddard T, McDonald AD, Alambeigi H, Kim AJ, Anderson BA. Unsafe bicyclist overtaking behavior in a simulated driving task: The role of

implicit and explicit attitudes. Accident Analysis and Prevention. 2020;**144**:1-11

[31] Farah H, Piccinini GB, Itoh M, Dozza M. Modelling overtaking strategy and lateral distance in car-to-cyclist overtaking on rural roads: A driving simulator experiment. Transportation Research Part F. 2019;**63**:226-239

[32] Rasch A, Moll S, López G, García A, Dozza M. Drivers' and cyclists' safety perceptions in overtaking ma-neuvers. Transportation Research Part F: Traffic Psychology and Behaviour. 2022;**84**:165-176

Chapter 6

Digital Electrohydraulic Braking Systems in Automotive Engineering

Nicolae Vasiliu, Andrei Rădulescu-Gamaleț,
Georgiana Claudia Vasiliu and Marius Daniel Bontoș

Abstract

The paper contains a review of the main world achievements in the field of digital electrohydraulic automotive braking systems and a short presentation of some theoretical and experimental studies performed by the authors to assess the static and dynamic behavior of recent and new braking systems. Design, modeling, digital simulation and experimental identification are used to point out the degree of mechanical, electrical and hydraulic complexity of some new achievements in the field.

Keywords: digital electrohydraulic brake servosystems, vacuumatic brakes, electromechanic brakes, hybrid electrohydraulic brakes, modeling, simulation, validation

1. Introduction

In 1980 BOSCH Company patented the first antilock braking system (ABS) system for automotive braking systems, generating wide research in all the important car design companies for reducing the mass of the device and add new functions, such as driver assistance for different roads, weather events, traffic conditions, etc. The most important design tool became the Real-Time Simulation concept, performed with different languages implemented in a proprietary-designed test bench. The main simulation languages used for this aim were NI Matrixx and LabVIEW, produced by NI Corporation [1]; Simcenter Amesim [2], developed by Siemens Corporation [3]; and SIMULINK, developed by The MathWorks Inc. [4], as an extension of MATLAB.

The next important progress in increasing automotive braking security was achieved by CONTINENTAL Company [5–7], which introduced the first hybrid input system in the flow control valve of the vacuumatic booster Ate Mk 25 (**Figure 1**). Moreover, for safety reasons, two vacuum chambers are used. The position of the second membrane is used for activating the emergency electromechanical brakes with the aid of a linear potentiometer.

The scheme of the ABS/TCS system (**Figure 2**) includes the force booster assembly **1**, the four calipers **2** and the electrohydraulic digital control unit **3**.

IntechOpen

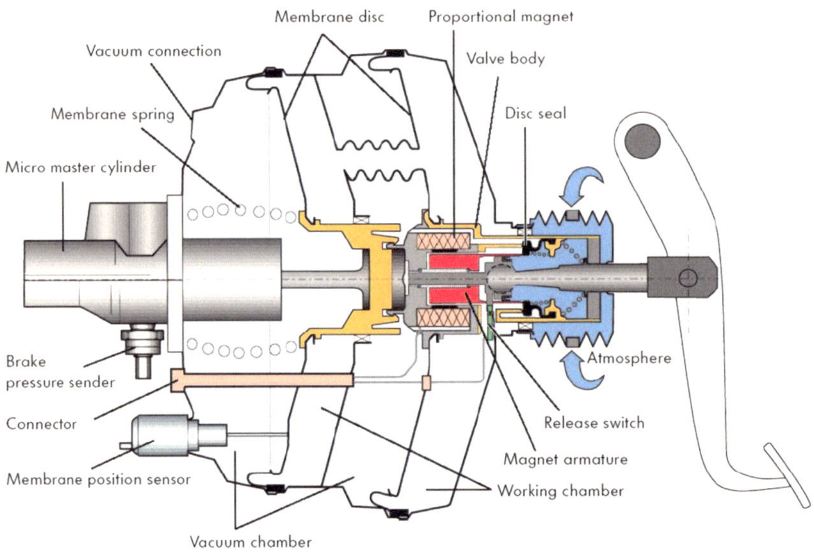

Figure 1.
Main cut view of the first hybrid electrohydraulic brake [5–7].

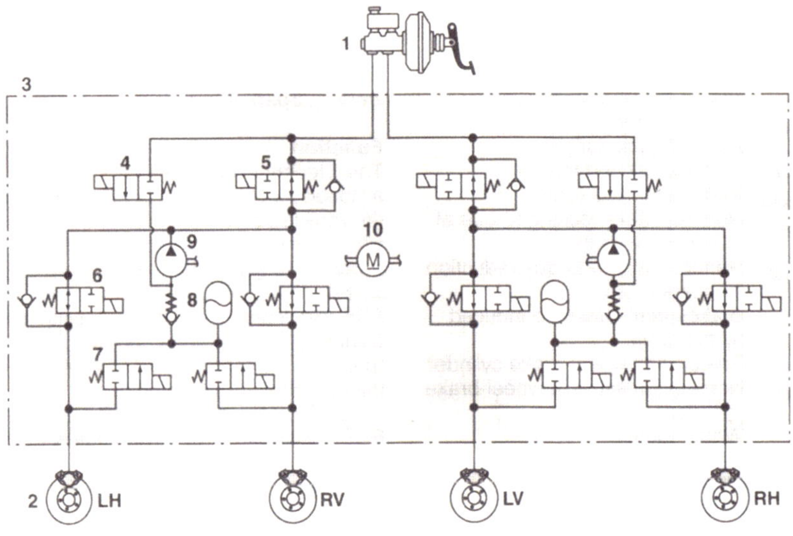

Figure 2.
The ABS/TCS hydraulic circuit for X brake circuit configuration.

The simplest Simcenter Amesim model of an ABS global system (**Figure 3**) allows a good prediction of the pressure variations in all active calipers (**Figure 4**) [8, 9]. A long series of dynamic braking and steering tests were successfully performed by the author's team with a VOLVO XC90 car. A complex braking system includes many additional components to help the driver to avoid any accident.

For example, in the studied case, all the attempts by the authors to create dangerous steering and braking situations were rejected by the electronic stability program (ESP) together with the ABS system (**Figures 5** and **6**). The ABS system response,

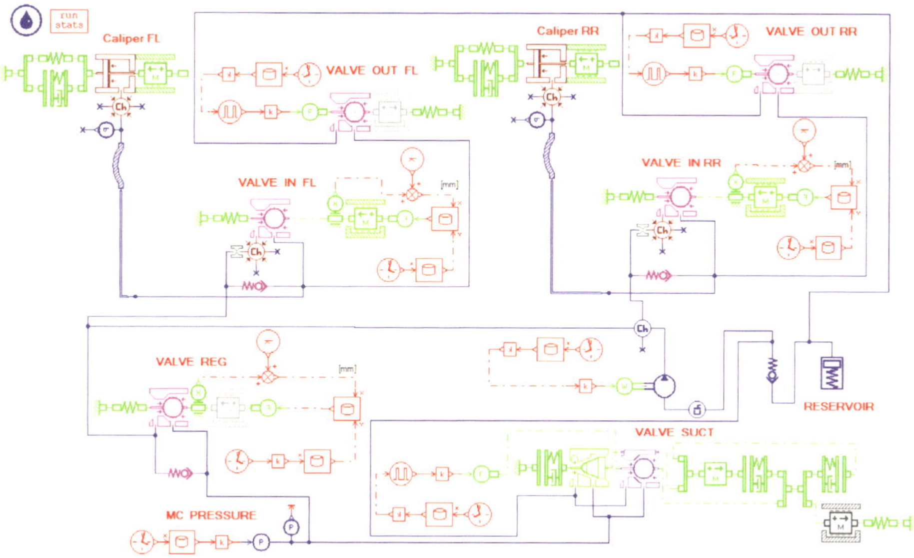

Figure 3.
Model of the ABS global system (standard system).

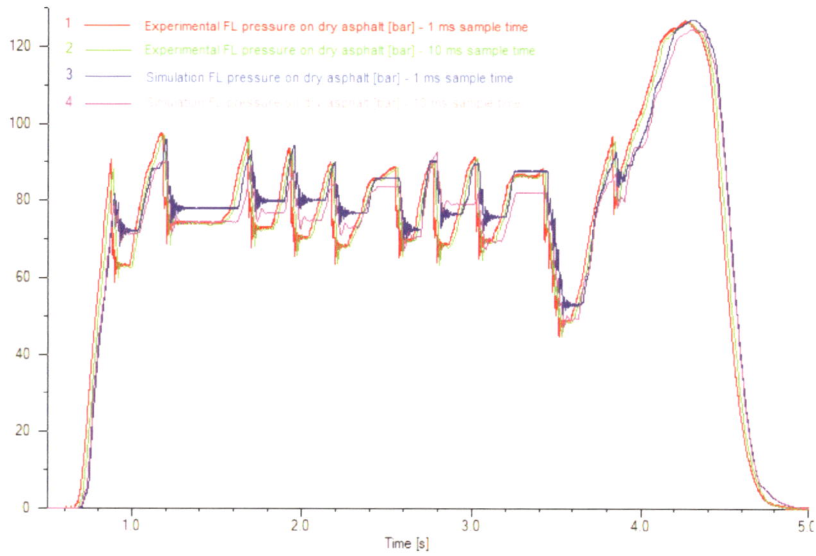

Figure 4.
Comparison between simulated and real pressures for front-left (FL) caliper on dry asphalt with 1 ms or 10 ms sample time (standard ABS system).

regarded by the pressure applied to the calipers, is shown in **Figure 7** for an emergency brake. The hydraulic systems act by successive pressure pulses of 82 Hz.

The concentration of all the piezoelectric accelerometers in a single robust sensor (sited in the car center) offers a proper set of simultaneous information. Consequently, this type of brake booster is still used for economic reasons in hybrid cars with turbo compressors, connected to a small electrically driven vacuum pump (**Figure 8**) [10].

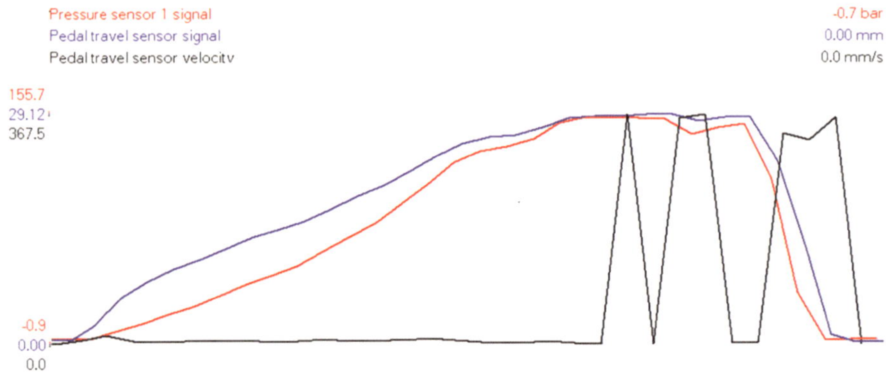

Figure 5.
Brake pressure, pedal travel, and velocity during a normal braking action.

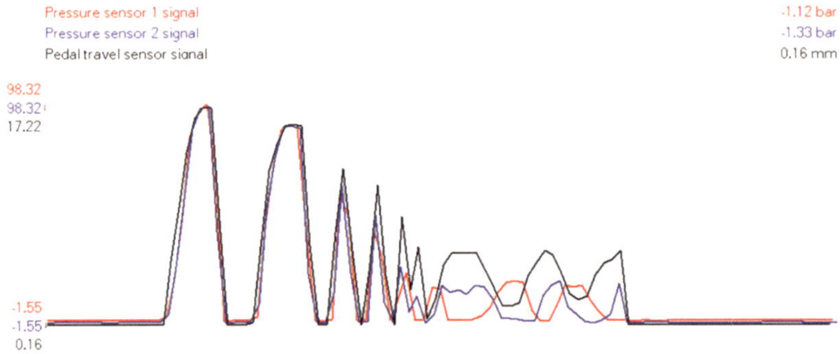

Figure 6.
Braking pressure (in red) and the brake pedal stroke during driving with over-steering.

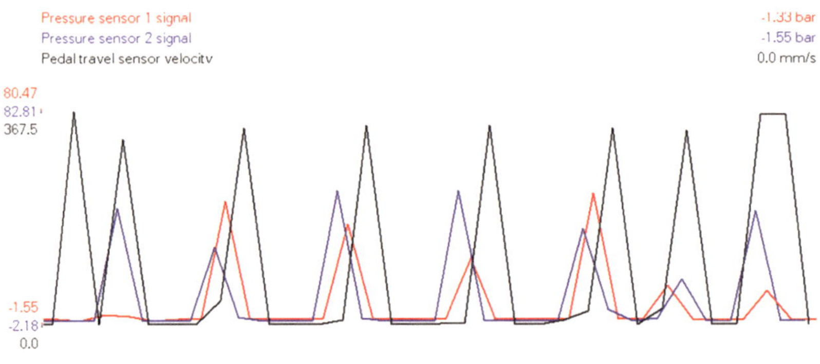

Figure 7.
Braking pressure, pedal travel, and velocity during an emergency brake.

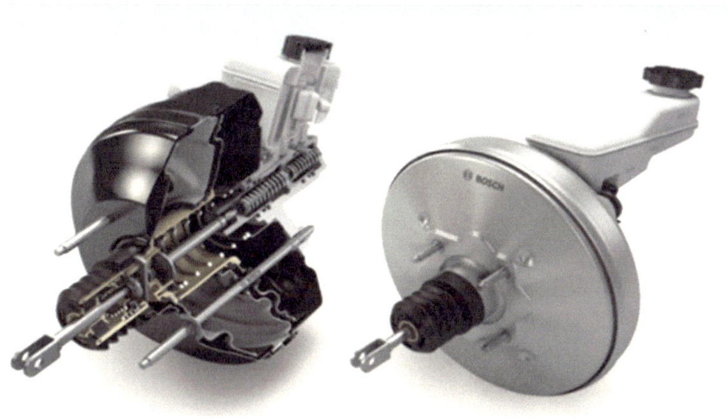

Figure 8.
Modern vacuumatic brake boosters (BOSCH).

2. Advanced ABS/ESC control systems

The field of driver assistance systems has been widely developed in the last decade. This has been made possible by technological progress, specifically in the development and production of electronic components. Shared use of subfunctions by various systems, information interchange between different systems, as well as various interdependencies are the logical consequences of this development. Electronic Stability Control can perform many other functions, such as Electronic Brake Force Distribution (EBD). To complete all the new functions, the ABS unit is completed with different sensors and CAN connections, keeping together a lot of components to avoid additional blocks or external digital connections. The scheme of the basic ABS system from **Figure 9** includes the components from **Figures 10** and **11** assembled in a compact and technological manner [11].

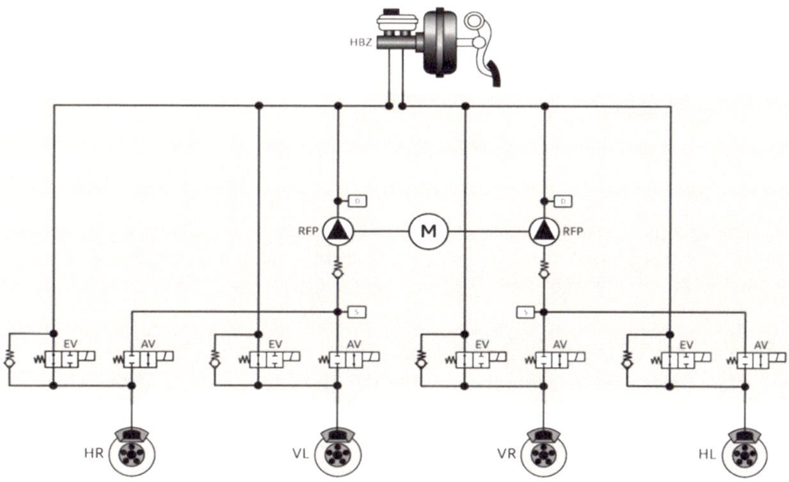

Figure 9.
Basic ABS module scheme.

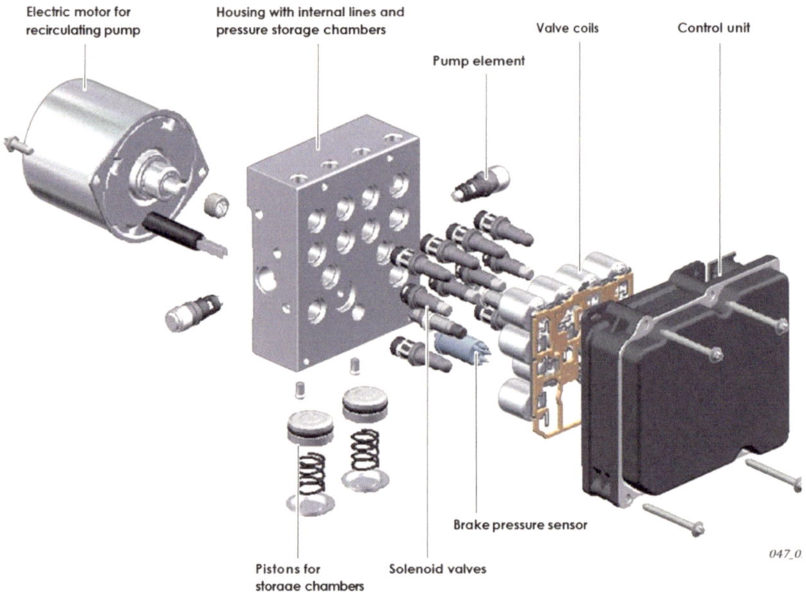

Figure 10.
Expanded view of an ABS hydraulic unit.

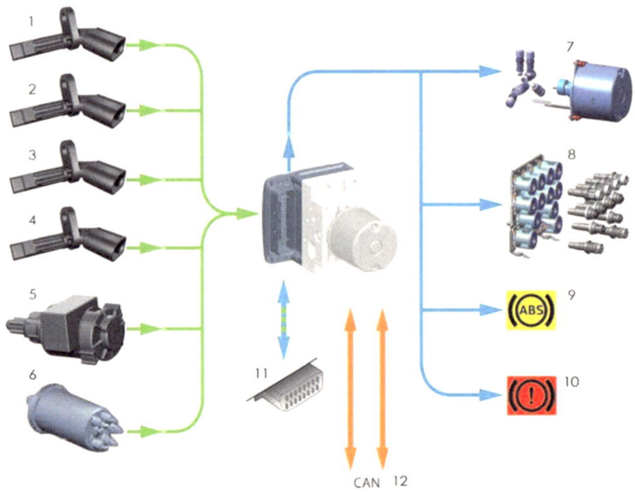

Figure 11.
The components of an ABS system: 1–4—speed sensors; 5—brake light switch; 6—brake pressure sender; 7—recirculating electric pump; 8—ABS warning lamp; 9—ABS warning lamp; 10—brke system warning lamp; 11—diagnostic port; 12—digital communications.

The three phases of the braking process are the following: (a) Brake pressure build-up by the driver (without active control); (b) "Brake pressure holding" function; and (c) "Brake pressure reduction" function.

ABS-controlled braking involves deploying the brake pressure reduction function, the brake pressure holding function and the brake pressure build-up function in rapid succession. The functions are also implemented by short, millisecond-duration pulses. The diagram from **Figure 12** shows such a sequence of pulses.

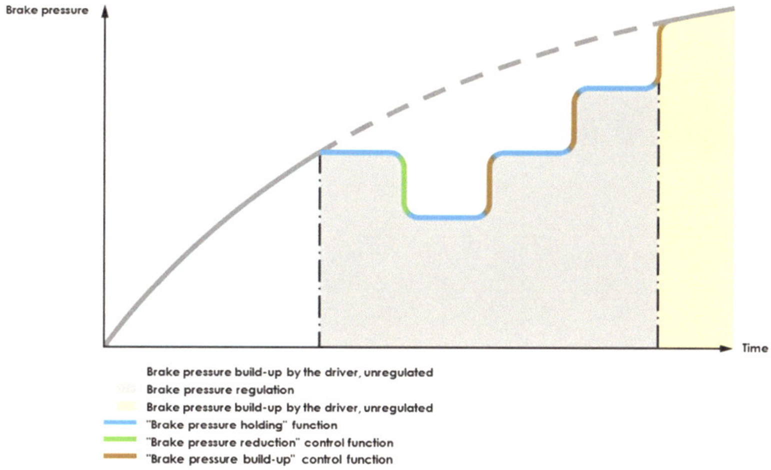

Figure 12.
ABS control function during a braking process.

The ABS system is usually combined with Electronic Stability Control (ESC) to keep the right direction of the car in special track conditions. The front right wheel on the vehicle shown here is braked while traveling in a straight line. The brake force F acting on this wheel is at a distance from the vehicle's vertical axis. As a result, the brake force produces a torque Md = F × S about the vehicle's vertical axis (**Figure 13**).

The ESC control software and the other control systems are located centrally in a control unit. Apart from the wheel speed sensors, other sensors that monitor the movements of the vehicle are required to implement the ESC function. For example, in current Audi models, the sensors for monitoring the movements of the vehicle are located in a central unit. In addition to ABS with Electronic Differential Lock (EDL), a steering angle sensor is also required to monitor the steering wheel angle. Warning lamps and displays notify the driver of the system's status. Depending on the model, the driver can shut down and change over certain systems using a switch (button).

As the name indicates, the function of ESC is to stabilize the vehicle in dynamically critical driving situations. The system uses three different control strategies for this purpose: active brake pressure build-up at certain wheels, intervention in the engine management system to reduce power and intervention in the gearbox control unit to suppress gearshifts. In vehicles with dynamic steering, the ESC also makes stabilizing steering angle corrections, if necessary. The focus of ESC is on braking

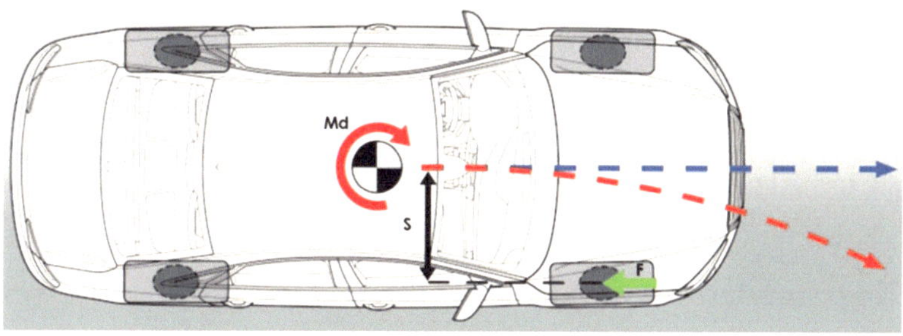

Figure 13.
The loss of stability during an asymmetrical braking event.

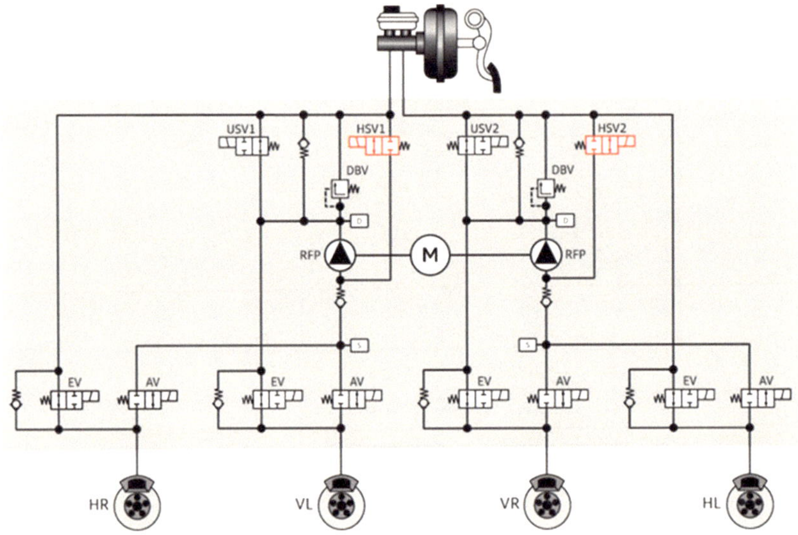

Figure 14.
ESC module hydraulic scheme.

intervention using the hydraulic scheme from **Figure 14**. The pressure shocks generated by the recirculating pumps are avoided by two storage chambers sited on the suction lines (S), and by two damper chambers placed on the delivery lines of the pumps (D). The fine-tuning of such a complex system by simulation needs a realistic assessment of all the small volumes of fluid from the control unit.

The ESC control software is integrated into a control unit together with the software for the ABS, EBD, EDL and TCS control units. The control unit determines and continuously compares the vehicle's actual behavior and nominal behavior. If the deviation between nominal behavior and actual behavior exceeds preset limits, then the control function will be activated. The measured values for yaw rate (yaw torque about the vehicle's vertical axis), and longitudinal and lateral acceleration are evaluated. The wheel slip values as well as driving speed, acceleration and deceleration are determined from the measurement data collected by the speed sensor. The brake pressure senders deliver information on the actual pressure in the brake system's primary circuit. In vehicles with an automatic gearbox, the ABS/ESC control unit also indicates by data bus message the gear that is currently selected. To determine the vehicle's nominal behavior, the following driver activities must be monitored: steering, accelerator pedal movements (acceleration/deceleration) and braking. The measured value provided by the steering angle sender indicates the direction in which the driver wants to go. When the driver operates the accelerator pedal, this is indicated to the ABS/ESC control unit by the engine control unit by a bus message. The signal generated by the brake light switch indicates to the control unit that a brake operation has been initiated by the driver. The measured value of the brake pressure sender delivers a redundant signal, which also serves as the basis for calculating stabilizing ESC braking intervention.

3. Modeling, simulation and test of the electromechanical brake servo

The iBooster [12, 13] created by BOSCH Company (**Figures 15** and **16**) is particularly suited to hybrid and electric vehicles. The actuation of the brake pedal generates

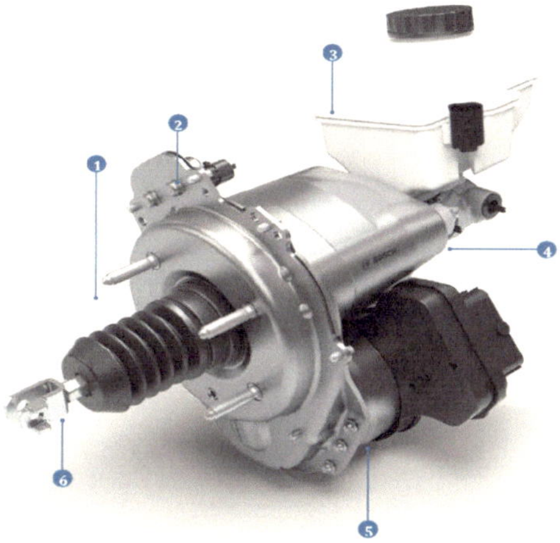

Figure 15.
Overall view of the second generation of the iBooster: 1—vehicle interface; 2—differential travel sensor; 3—reservoir with the level sensor; 4—tandem master cylinder; 5—power pack (brushless motor and electronic control unit); 6—input road.

Figure 16.
Cut view of the second generation of iBooster.

the control signals for an electric motor, while a three-stage gear unit converts the torque of the motor into the necessary boost force.

The required braking pressure can be built electromechanically three times more quickly than the previous vacuumatic systems and adjusted with great accuracy through a classical digital control system. In combination with ABS, ESP, etc., the iBooster provides the good redundancy required by automated vehicles. The prediction of the performances of such a hybrid system can be performed by Simcenter Amesim basic simulation network from **Figure 17** [14].

A common duty cycle of the system, introduced by the pedal travel sensor is specified in **Figure 18**. The results of the simulations are presented in **Figures 19–24**.

The very low inertia of the brushless motor allows a very low response time (about 0.01 ms), for a nominal power of 370 W supplied by the car's 12 V battery (**Figure 19**).

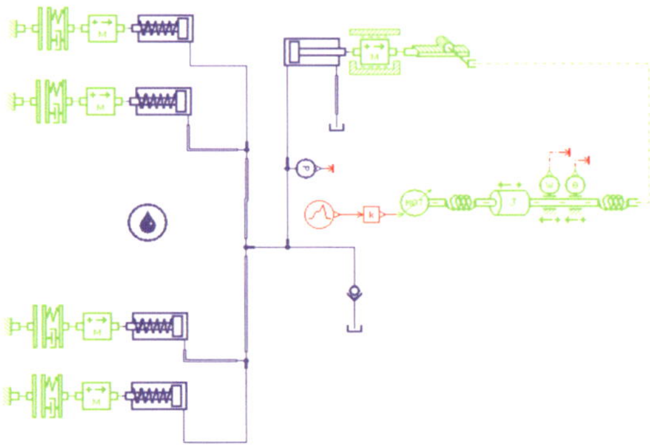

Figure 17.
Simulation network of iBooster for a normal braking procedure.

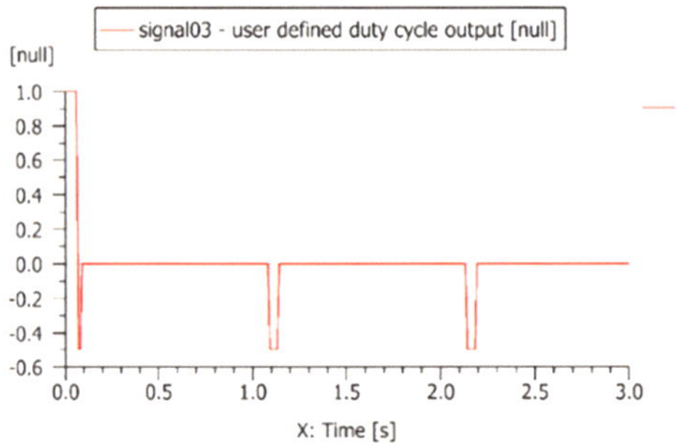

Figure 18.
The duty cycle output is introduced by the pedal travel sensor.

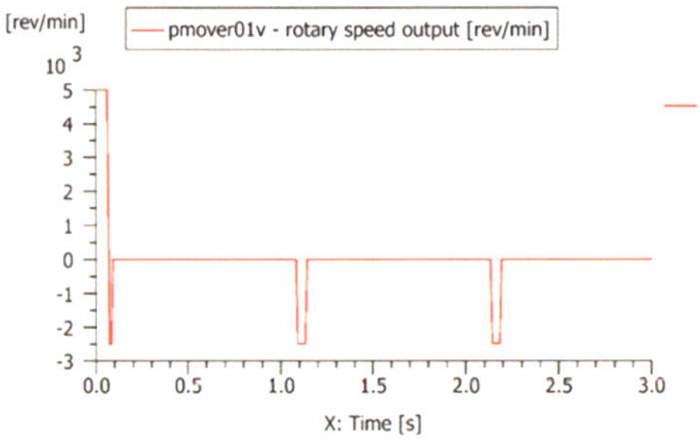

Figure 19.
Brushless motor rotary speed variation.

4. Experimental validation of the simulation model

The understanding of a complete brake process needs a complex investigation. The presented ones were globally validated by some experimental research carried out on a full ID3 electric car, running on a good quality highway during a few quasi-identical driving cycles. The diagnosis system used to obtain real-time results for the car was VCDS ver.23.1.0 (2023 MY update) from ROSS-TECH Company [15]. The diagrams in **Figures 25–28** were obtained by periodic cycles of quick accelerations and sudden decelerations. They confirm the validity of the numerical simulations.

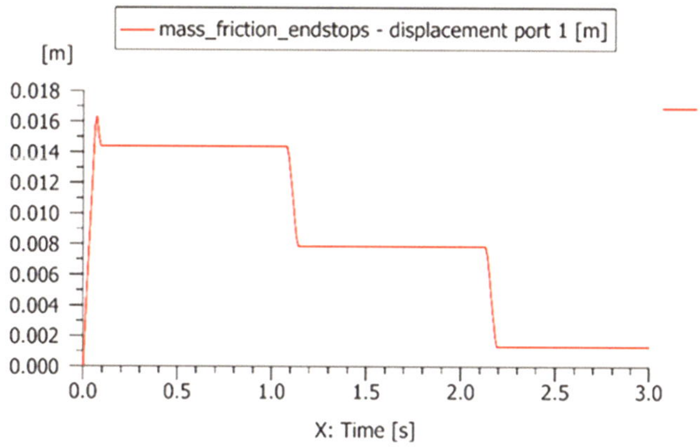

Figure 20.
Displacement of the pump piston actuated by the brushless motor.

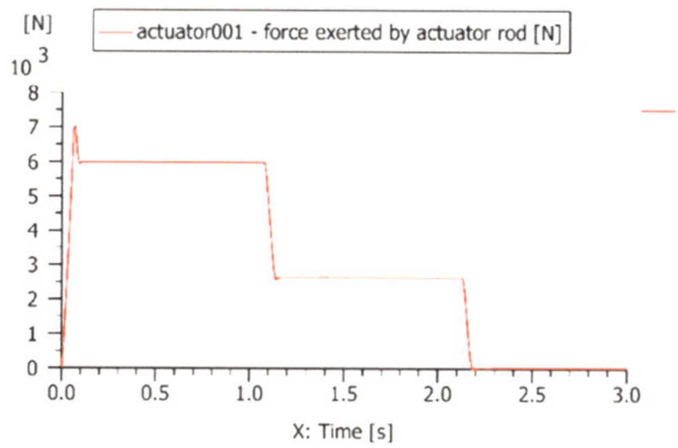

Figure 21.
Force applied to the pump piston.

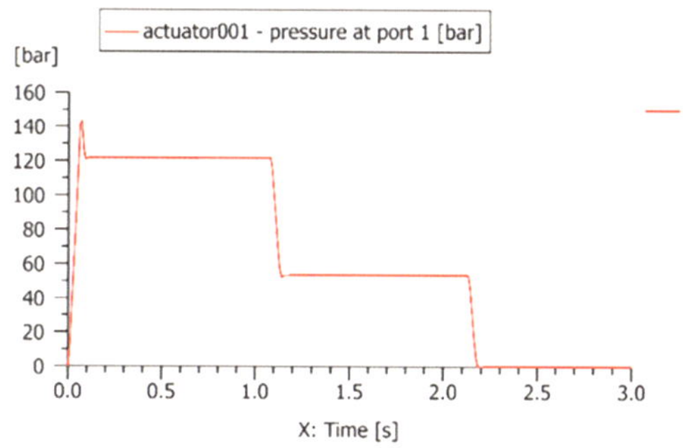

Figure 22.
Electromechanical pump pressure variation.

The frequent velocity car variations during the test performed in harsh conditions on a highway demand a comfortable (simulated) braking effort for the driver. Compatibility with all ABS/ESC functions makes the electromechanical brake system very attractive for hybrid cars.

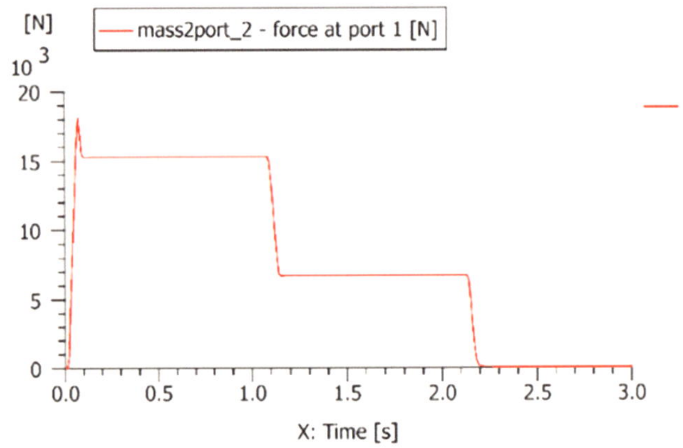

Figure 23.
Pressure force applied on the pistons of a fixed caliper.

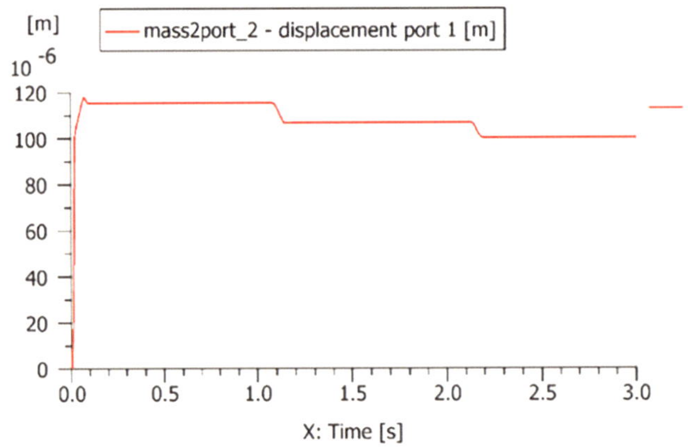

Figure 24.
Typical axial displacement of the biggest caliper piston.

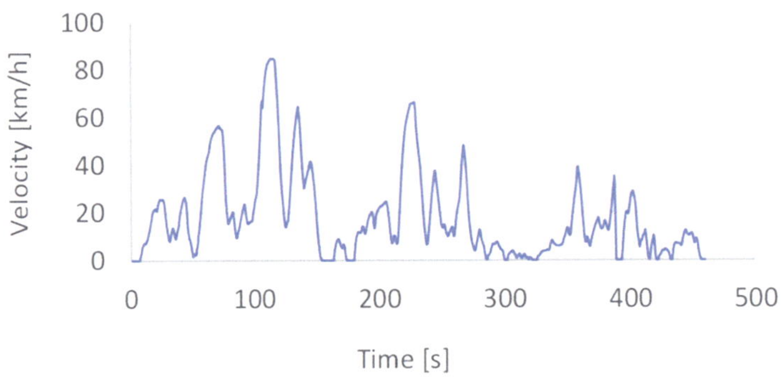

Figure 25.
Car velocity variation in the first test cycle (460 s).

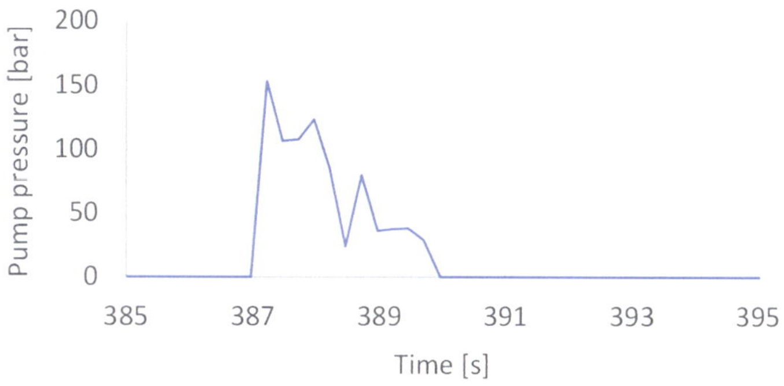

Figure 26.
The pump pressure variation during an emergency brake at 387 s after the beginning of the test (with ABS activated for 3 s).

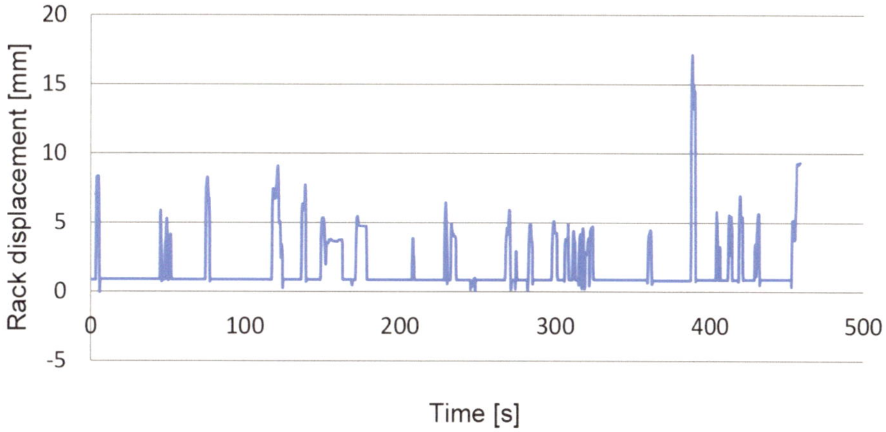

Figure 27.
The pump rack displacement during the first test cycle, with a maximum at the second 387 (ABS is activated).

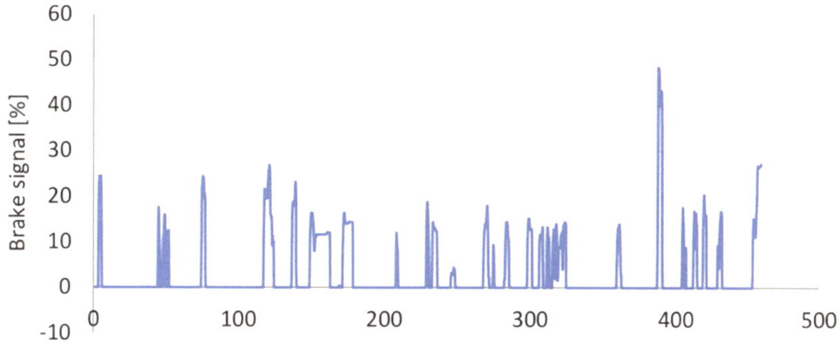

Figure 28.
The variation of the relative brake input signal during the test (max. value: 48%).

5. Modeling, simulation and test of the electrohydraulic brake servo

The wide interest in driving electric cars offered the electrohydraulic servosystems designers a new development challenge. The first electrohydraulic brake-by-wire system named MK C1 was produced by CONTINENTAL AG in 2016, and it was successfully implemented on Audi e-TRON E55 in 2018 [16]. It provides the driver with a brake pedal feeling that remains constant, even when recovery is taking place. The operation principle results from **Figure 29**.

BOSCH Company developed similar concepts [15, 17–21]. The new generation of Audi Q6 e-tron (type F4) which will be produced starting in 2025 includes similar complex equipment with some peculiarities regarding the pedal force simulator unit, and the progressive control of the brake release by four very small two-way servovalves (**Figure 30**). The set of functions of this new equipment is presented in **Figure 31**. The two operating regimes (fully electric and hybrid) are presented in **Figures 32** and **33**.

The numerical simulation of the electrical duty cycle of the system [22–24] shows a good dynamic for any duty cycle output introduced by the pedal travel sensor (**Figures 34–38**).

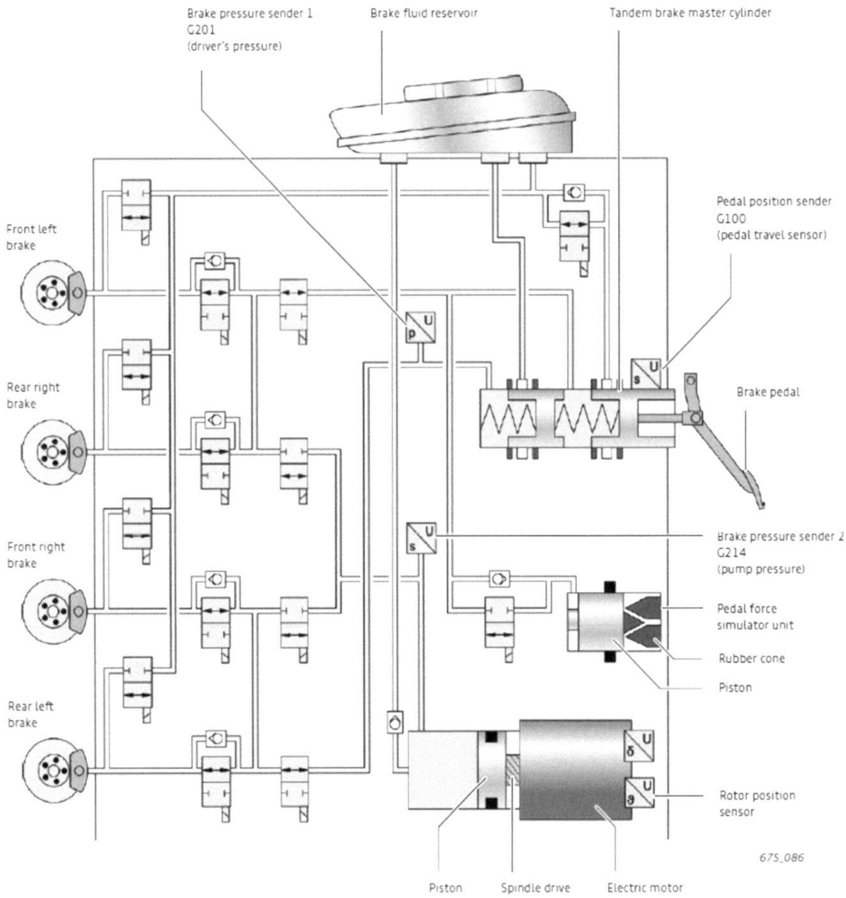

Figure 29.
The brake regulation system MK C1.

DOI: http://dx.doi.org/10.5772/intechopen.1009352

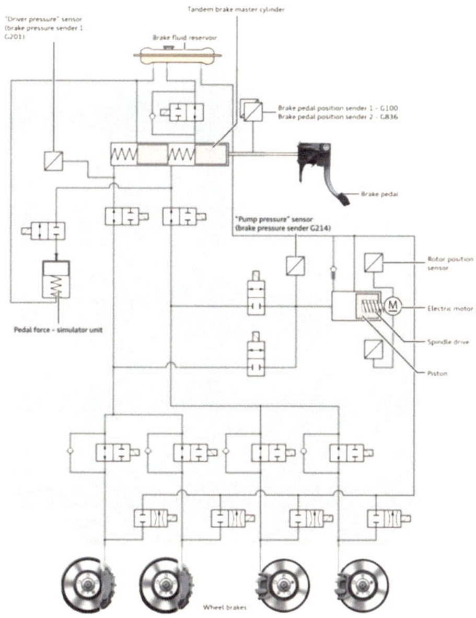

Figure 30.
The scheme of the BOSCH IPB 1.1 integrated brake control system.

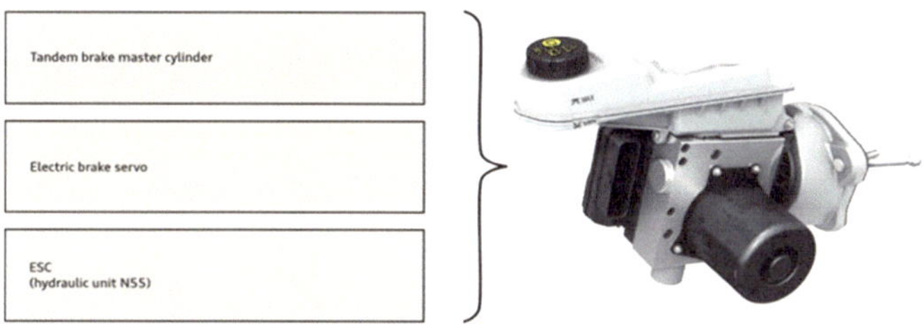

Figure 31.
BOSCH IPB 1.1 integrated brake control system.

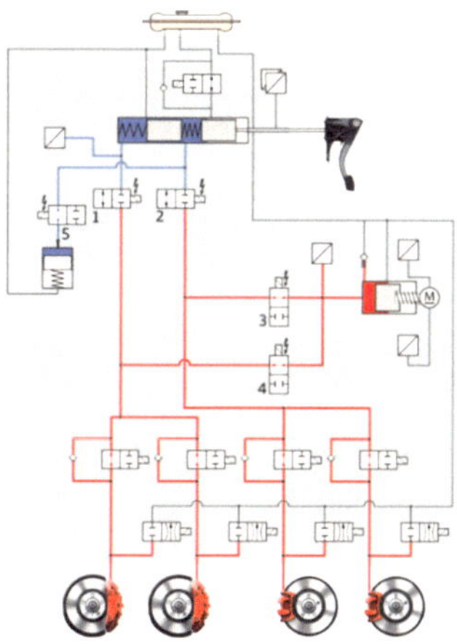

Figure 32.
Full electric operation mode of the BOSCH EHB with pedal force simulator unit activated.

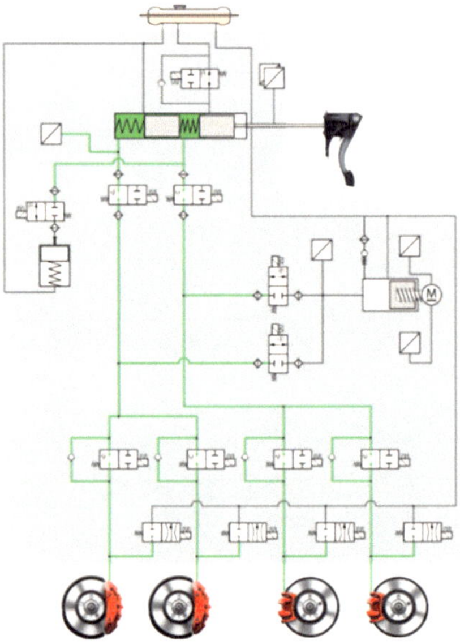

Figure 33.
Hybrid operation scheme of the BOSCH EHB without force simulator unit activated.

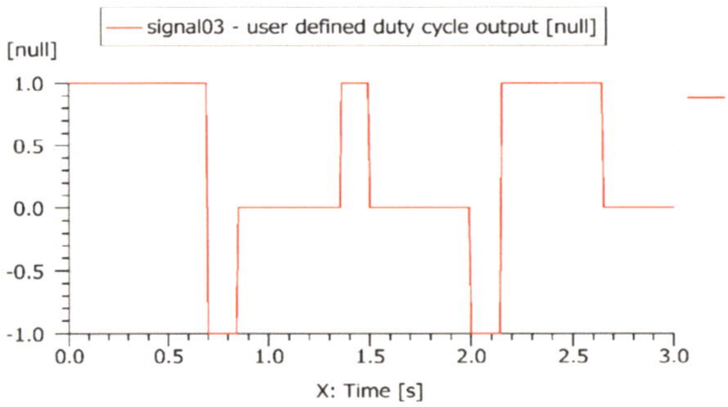

Figure 34.
The duty cycle output introduced by the pedal travel sensor.

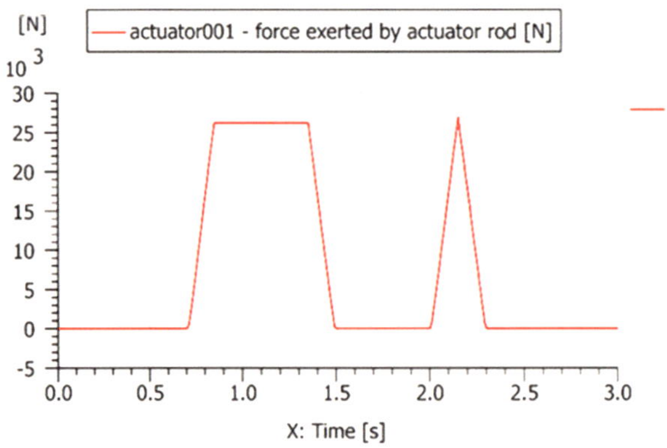

Figure 35.
The force exerted by the electromechanical actuator on the pump piston.

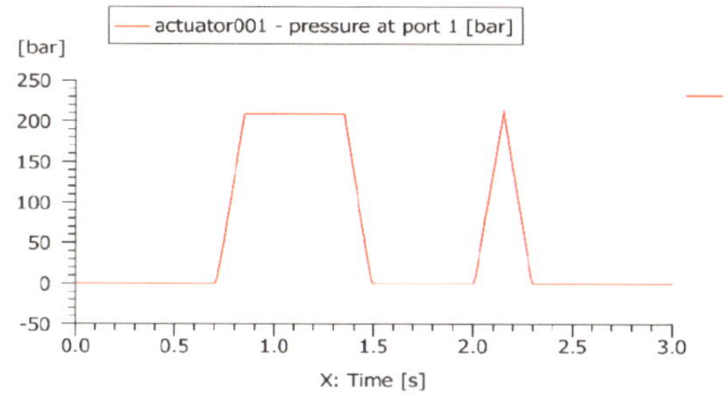

Figure 36.
Pressure variation in the delivery port of the electromechanical pump.

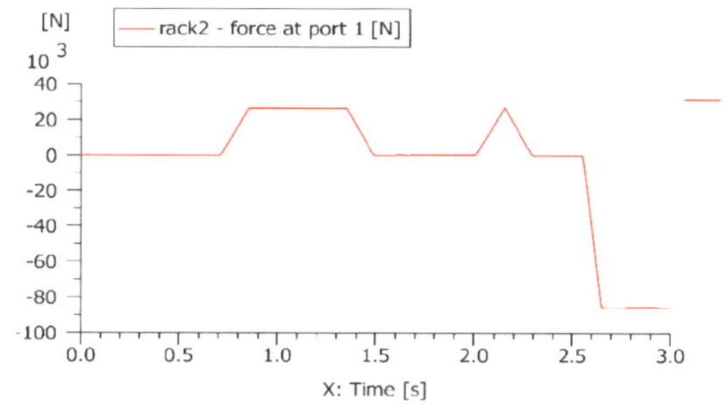

Figure 37.
The pressure force is applied on the pistons of a fixed caliper of the front axle (six pistons of 30-36-38 mm).

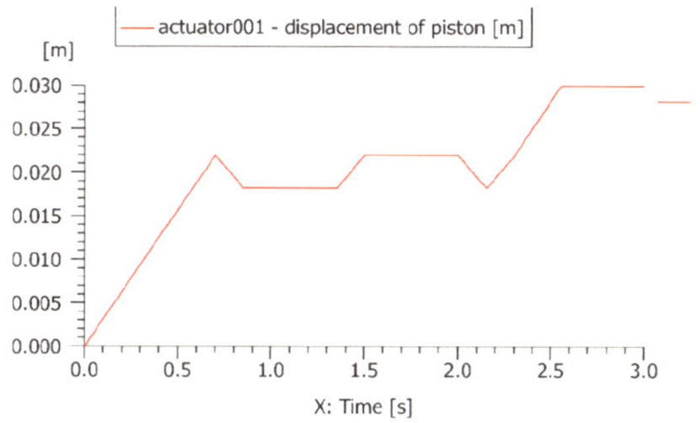

Figure 38.
Typical axial displacement of the biggest caliper piston (max. 30 μm).

6. Experimental validation of the simulation model

The above numerical simulations were validated by experimental research carried out on a modern electric car (AUDI e-tron 55) running in harsh traffic conditions. The diagrams from **Figures 39–42**, obtained by sudden braking cycles, prove the accuracy of the predictions offered by simulations [22–24].

The virtual pressure felt by the driver is multiplied by 7.6 times without starting the ABS system.

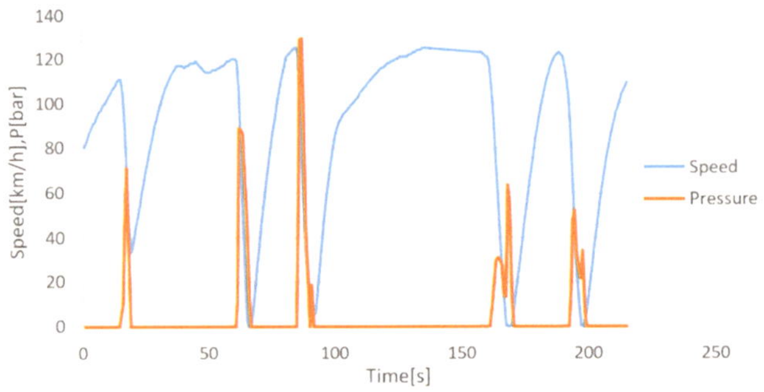

Figure 39.
A typical accelerating-braking test was performed on the highway.

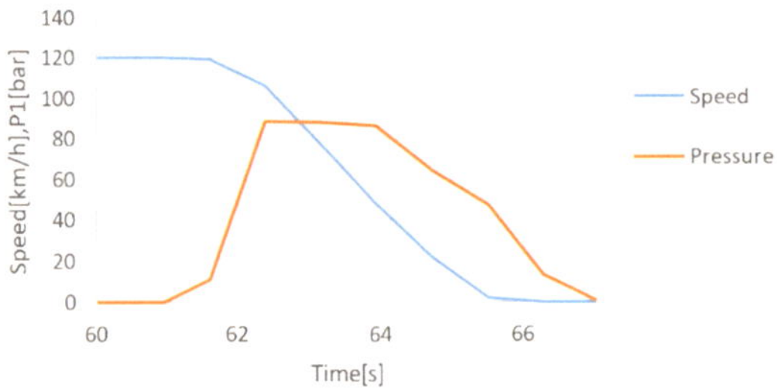

Figure 40.
Velocity reduction during an emergency brake.

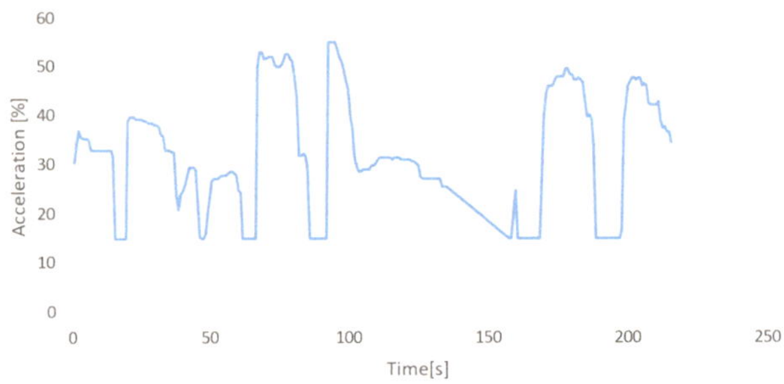

Figure 41.
Variation of the car acceleration in a typical test cycle.

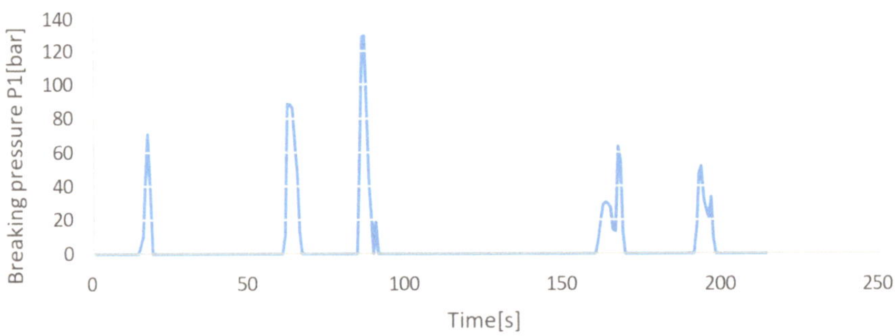

Figure 42.
Variation of the braking pressure from calipers.

7. Conclusions

The high progress achieved in the design of digital electrohydraulic braking systems [16, 25] can improve the security level of automotive systems. The use of Real-Time Simulation with Hardware-in-the Loop methodology will reduce the design preliminary phase duration, in close connection with the realistic technical specifications of all the braking systems components. The hybrid training (in hardware and software) of the automotive design staff and students has become an important component of the progress in this important field of economy and society [26, 27].

Acknowledgements

We are particularly grateful to Dr. Jan Leuridan, former CEO of Siemens Industry Software NV, and Dr. Cristi Irimia, General Manager of Siemens Industry Software SRL. For 15 years, they supplied us with professional technical reports and high-level software for teaching our students and creating new industrial hydraulic control systems.

Author details

Nicolae Vasiliu*, Andrei Rădulescu-Gamaleț, Georgiana Claudia Vasiliu and Marius Daniel Bontoș
University Politehnica of Bucharest, Romania

*Address all correspondence to: vasiliu1946@gmail.com

IntechOpen

© 2025 The Author(s). Licensee IntechOpen. This chapter is distributed under the terms of the Creative Commons Attribution License (http://creativecommons.org/licenses/by/4.0), which permits unrestricted use, distribution, and reproduction in any medium, provided the original work is properly cited.

References

[1] LabVIEW Fundamentals, Updated 2024. Available from: https://www.ni.com/docs/en-US/bundle/labview/page/user-manual-welcome.html

[2] Lebrun M, Richard CL. How to create good models without writing a single line of code. In: The Fifth Scandinavian International Conference on Fluid, SICFP'97; 28-30 May 1997; Linköping, Sweden. ISBN 9178719755, 9789178719754

[3] Vasiliu N, Vasiliu D, Călinoiu C, Puhalschi R. Simulation of the Fluid Power Systems with Simcenter Amesim. Boca Raton, FL, USA: CRC Press, Taylor & Francis; 2018

[4] Learning SIMULINK. The MathWorks Inc.; 2004. ISBN 0-9755787-3-1. Updated 2024. Available from: https://www.mathworks.com/help/simulink/release-notes.html

[5] Continental Teves New Generation of Braking Systems. 2000

[6] Dardac L. Research on the automotive electrohydraulic brakes [master thesis]. Bucharest, Romania: University POLITECHNICA of Bucharest; 2011

[7] Dardac L, Vasiliu N, Vasiliu D. Experimental assessment of an ABS system, U.P.B. Scientific Bulletin Series D. 2012;**74**(3):195-208. ISSN 1454-2358

[8] Dragne FD, Alirand M, Oprean IM, Vasiliu N. ABS valve model reduction by AMESim. Proceedings of the Romanian Academy, Series A: Mathematics, Physics, Technical Sciences, Information Science. 2009;**10**(2):189-196. ISSN 1454-9069

[9] Dragne FD. Researches on numerical modeling and simulation of automotive hydraulic systems with application on an ABS/ESP braking system [PhD thesis]. Bucharest, Romania: University POLITEHNICA of Bucharest; 2012

[10] Robert Bosch GmbH. Automotive Handbook. 11th ed. Karlsruhe: John Wiley & Sons; 2022. ISBN 978-1-119-91190-6

[11] Audi ESC systems. Self-Study Programme 475, AUDI AG D-85045, Ingolstadt Technical status 07/12

[12] Robert Bosch GmbH, iBooster 2nd Generation, Vacuum-independent Brake Booster; 2017

[13] Audi Q7 e-tron quattro (type 4M), Self Study Programme 649, AUDI AG I/VK-35. 2015

[14] Rădulescu-Gămăleț AC, Nica OI, Călinoiu C, Vasiliu N. Electromechanical braking servos for for electric vehicles – Simulations and experimental validation, U.P.B. Scientific Bulletin Series C. 2023;**89**(3):293-308. ISSN 2286-3541

[15] Advanced Breaking Technology. Failsafe Workshop Manual. Western Australia: Toyota LANDCRUISER, Perth Digital Edge; 2021

[16] Pinkow S., The Future of Brakes-Brake Systems of the Future, Continental-Automotive. Available from: https://www.continental.com/en/press/studies

[17] Toyota Hybrid System THS II, Toyota Motor Corporation, Public Affairs Division 4-8 Koraku 1-chome; May 2003; Bunkyo, Tokyo, 112-8701, Japan

[18] McDonald D. Electric vehicle drive simulation with MATLAB/Simulink. In: Proceedings of the North – Central

Section Conference; 24 March; Ohio Northern University. American Society for Engineering; 2012

[19] Enang W, Bannister C. Modelling and control of hybrid electric vehicles (a comprehensive review). Renewable and Sustainable Energy Reviews. 2017;**74**:1210-1239

[20] Alexander A, Vacca A. Longitudinal vehicle dynamics model for construction machines with experimental validation. International Journal of Automotive and Mechanical Engineering;**14**(4):4616-4633. ISSN: 2229-8649 (Print); ISSN: 2180-1606 (Online), December 2017©University Malaysia Pahang Publishing. DOI: 10.15282/ijame.14.4.2017

[21] Fraj A, Budinger M, Halabi TE, Mare J-C, Negoita GC. Modelling approaches for the simulation-based preliminary design and optimization of electromechanical and hydraulic actuation systems. In: AIAA/ASME/ASCE/AHS/ASC Structures, Structural Dynamics and Materials Conference. Honolulu; April 2012. DOI: 10.2514/6.2012-1523

[22] Rădulescu-Gămăleț AC, Vasiliu D, Călinoiu C, Cojocaru-Greblea T, Nica OI. Electrohydraulic braking systems for electric and hybrid vehicles, U.P.B. Scientific Bulletin Series C. 2022;**85**(3):419-432. ISSN 2286-3541

[23] Rădulescu-Gamaleț AC. Researches on the hybrid and electric transmissions [Ph.D. thesis]. Romania: University POLITEHNICA of Bucharest; 2024

[24] Cojocaru-Greblea T. Researches for improving the efficiency of the electrohydraulic servomechanisms [Ph.D. thesis]. Romania: University POLITEHNICA of Bucharest; 2022

[25] SENSIFY-The New Standard in Braking, Brembo N.V. 2024. Available from: https://www.brembo.com/en/company/news/future-brake

[26] Audi e-tron (type GE) Self-study programme 675, AUDI AG I/VK-35. 2021

[27] Audi Q4 e-tron (type F4) Self-study programme SSP 685, Ingolstadt, 02/2021

Chapter 7

Enhanced Object Recognition in Autonomous Driving through LiDAR Point Cloud Data Compression

Yanjun Zhang and Xiaoyu Sun

Abstract

In autonomous driving, lidar-derived point cloud data is crucial for detecting objects and understanding the vehicle's environment. However, the large volume of raw data poses challenges in storage, transmission, and real-time processing. Compression of lidar point cloud data is vital, as it reduces data size without significantly losing geometric and semantic information, thus enhancing autonomous driving system performance. Compressed data enables swift and accurate object recognition, improving safety and reliability. This chapter examines current compression methodologies, their strengths, limitations, and applications. It focuses on vehicle recognition as a practical application, detailing the chosen algorithm's processing and compression of lidar data. The impact of compression on vehicle recognition accuracy and efficiency is discussed, with implications for intelligent transportation systems and future advanced compression techniques for real-time data processing in autonomous vehicles.

Keywords: autonomous driving, boundary extraction, point cloud data compression, three-dimensional point cloud sequence data coding and decoding, adaptive bit-rate coding and decoding

1. Introduction

With the development of computer technology, the digitization of complex real-world scenes has become a reality. In the field of surveying and mapping, concepts such as "Smart City" and "Digital Earth" are receiving increasing attention. Constructing large-scale and complex virtual scenes requires a large amount of three-dimensional coordinate data of target objects that can be directly processed by computers. General technical means such as cameras and flat scanners can only obtain two-dimensional coordinate information of targets, which cannot meet the requirements. At the same time, traditional surveying and mapping methods such as image back-forward intersection and broken-step surveying can directly obtain the three-dimensional coordinate information of ground objects, but they have a long working cycle and a low point density. The three-dimensional lidar system adopts an active imaging method and can directly obtain high-precision and high-density three-dimensional coordinate information on the surface of target objects. Its intelligent and automated data acquisition characteristics provide powerful technical support for the digitization of large-scale and complex scenes, making it possible to obtain spatial information at high speed and high density.

Vehicle-mounted lidar systems, with their characteristics of high precision, high density, and all-weather operation, are widely used in road detection, vehicle detection, urban three-dimensional modeling, and other aspects. However, at the same time, the scene point cloud data they acquire is huge, reaching hundreds of millions within a short distance. With the development of lidar technology, the order of magnitude of point clouds is constantly increasing, while the corresponding methods for storing, processing, and transmitting massive point cloud data are progressing slowly. Therefore, the compression of point cloud data has become a key research area.

2. Research status of the compression of point cloud data

The following is a review of the research status: three-dimensional point cloud data compression, three-dimensional point cloud sequence data compression, and adaptive bit-rate compression technology.

2.1 Three-dimensional point cloud data compression

The algorithms for three-dimensional point cloud data encoding and decoding can be divided into the following four types: point cloud compression methods based on Level of Detail (LoD), point cloud compression methods based on clustering methods, point cloud compression methods based on graph transformation, and the recently emerging point cloud compression methods based on learning.

2.1.1 Point cloud compression methods based on LoD

First of all, the point cloud compression methods based on LoD can be further divided into those based on the point cloud data representation forms of binary tree, quadtree, and octree.

At present, 3D point cloud data represented by octree is receiving increasing attention. Octree representation is a form of 3D point cloud data, in which each node represents a 3D bounding box, and this bounding box is recursively decomposed into eight sub-leaf nodes.

Ding Caihong et al. [1] proposed a point cloud geometric prediction method based on neighborhood information. This method constructs an octree according to the predefined depth of the tree structure. It starts from a root node representing the entire space containing the point cloud. Then, at each layer of the octree, eight leaf nodes are generated as child nodes, and these child nodes correspond to the uniform subdivision of the parent 3D cell. A binary occupancy code is assigned to each child node, where a value of 1 indicates an occupied cell and 0 indicates an empty cell. The binary occupancy code can be directly entropy-coded in single-rate compression or used to find change points in further encoding schemes.

In other words, each layer of the octree can be interpreted as a Level of Detail (LoD) and is used to further subdivide the space. The construction process of the octree terminates when the user-defined maximum depth level is reached, which is equivalent to specifying a certain number of quantization bits for the geometric representation.

In addition, the octree model provides a stable three-dimensional spatial structure, which can evidently be used for motion estimation or compensation purposes in consecutive frames. The octree sequence t completely describes the octree representation used. The octree sequence t defines the binary pattern of the octree decomposition, starting from the root node (depth-first traversal of the tree).

Garcia et al. [2] proposed a typical octree-based lossless intra-frame compression method. The octree sequence t fully describes the octree representation used and defines the binary pattern of the octree decomposition. Another two sequences are used to generate a set of contexts: the parent value v, representing the binary value corresponding to its parent value, which indicates the occupancy rate, and the bit position p, which defines the ordered octree position of the corresponding point. The t sequence is finally conditionally encoded according to the context generated by the v and p sequences. Meanwhile, an arithmetic encoder and coding method [3] are used, and the context can be further applied to temporal geometric prediction.

In Ref. [4], the octree-based representation is applicable to streaming media applications. First, the point cloud of each frame in the streaming media data is decomposed into an octree. Then, the differences in the octree data structures of consecutive frames are calculated and encoded. The octree structures of two consecutive frames are respectively established and serialized into binary sequences representing occupancy codes. The change parts of the octree are detected and compared by using the exclusive disjunction operator. Finally, the obtained difference vectors are entropy-encoded. This method offers the advantages of differential octree representation, saving computation, and reducing memory requirements.

The lossless inter-frame compression method proposed in Ref. [5] also takes into account the temporal aspect of the 3D geometric sequence. The principle is to re-order the points in the octree structure of the current 3D frame according to the prediction of the previous frame before entropy encoding. Therefore, the temporal consistency of points with similar attributes and relatively low motion can be improved.

Other tree-based methods utilize k-d trees [6] or combine binary trees and quadtrees [7]. The method proposed in Ref. [6] is a binary tree-based method and is widely used for efficient neighborhood search for point geometry due to the lower computational complexity of binary trees. Unlike octree-based methods, the k-d tree divides the space in half at each internal node, which makes the k-d tree highly dependent on time-varying point cloud bounding boxes. Therefore, it is considered a particularly effective intra-frame encoding mode, but the construction of k-d trees is computationally expensive in the case of inter-frame encoding. Kathariya et al. [7] proposed the Binary Tree Quad Tree (BTQT) structure. It uses a binary tree to segment the point cloud after constructing a complete octree to detect empty blocks, avoiding encoding the full-depth octree. The binary tree blocks contain the local surface of the point cloud, and any local surface with planar features is projected onto a 2D plane and encoded using a quadtree. Other surfaces that are not flat enough are encoded using a full octree. The scalable encoding solution proposed by Kathariya et al. [7] can effectively compress point cloud data at variable rates, resulting in multiple Levels of Detail (LoD), which corresponds to the gradually increasing complexity of the three-dimensional representation. Ren yan et al. [8] optimized the issue of limited time savings in the encoding structure of the quadtree plus binary tree. By using a random forest model to train CU (Coding Unit) models of different depths, the number of times of traversing the partitioning modes and calculating the rate-distortion cost was reduced, and the encoding time was saved. Since adaptive bit rates and quality are in high demand for data transmission nowadays, this leads to the discussion of the importance of multiple LoD processing in this chapter.

LoD-based methods are usually based on or combined with the structured decomposition. In an early study, Huang et al. [9] proposed a general-purpose point cloud encoder based on the LoD structure of an octree. It estimates the statistical average values of the geometric center, normal, and color in each non-empty cell of the octree and represents the model part in each cell. To better utilize color correlations, Huang et al. [9] proposed a new method for determining an adaptive color box. This method uses principal component analysis to transform the coordinates in the RGB color space into a new point cloud frame where color correlations are more concentrated and align with the eigenvectors obtained from the principal component analysis. Similarly, in Ref. [10], the Generalized Lloyd Algorithm (GLA) is used to cluster points to generate a coarser LoD. This method can achieve adaptive color quantization using the Generalized Lloyd Algorithm (GLA).

Kitago et al. [11] proposed another LoD-based method, namely adaptive sub-sampling of point clouds. This method defines a cost function that integrates redundancy and non-uniformity information to prioritize these points. Then, the point set is sub-sampled according to the main geometric features of the model and the local sampling density. Subsequently, the problem is transformed into minimizing the coefficients of compactly supported radial basis functions [11, 12]. Finally, the position information and relative normal information of the points are used to estimate the surface, and the reconstructed points are re-scattered on the surface to reduce the distortion of the reconstructed points. By choosing different cost functions, multiple LoDs can be achieved with the help of the sub-sampling process considering different precisions of the reconstructed points.

In Ref. [13], a voxel-based, real-time, and adaptive Level of Detail (LoD) method is proposed, which takes into account network conditions and changes in application-related parameters. Consequently, the response time required by the application and the available network bandwidth can be used to dynamically calculate the target compression ratio. This method decomposes the point cloud into 2D images with height maps, and then applies JPEG image encoding for compression, achieving a balance between quality and compression ratio by controlling the resolution of the voxels.

The sparse voxel matrix is described as a voxelized point cloud in Ref. [14]. Voxels are obtained by decomposing the point cloud space into blocks, corresponding to the discretization of 3D space on a regular grid defined by a given step size (i.e., cell size). The voxelization process transforms the geometric coordinates of the point cloud into integer representations. Multiple LoDs are generated by changing the step size selected in the voxelization process. The graph-based motion estimation method proposed in Ref. [15] and the motion-compensation compression technique proposed by Queiroz et al. [14] show high efficiency, ensuring real-time processing of voxelized point clouds. In this case, 2D blocks are replaced by 3D cubic blocks called "voxels". The matching process of consecutive frames determines the correspondence between voxels by minimizing the distance between geometric and color information. The geometric residuals are regarded as motion vectors and associated with each voxel or voxel cube, and the color residuals are regarded as color prediction errors.

The method in Ref. [14] yields better color estimation, yet it introduces some noise caused by geometric displacement. Therefore, in the optimization process, further research can be conducted on how to select a reasonable parameter to balance geometric and color information.

2.1.2 Point cloud compression methods based on clustering

Since dense 3D point clouds can be segmented or clustered into different parts based on color or motion information, the point cloud compression methods based on clustering draw on the clustering principles in machine vision. Zhang et al. [16] proposed a point cloud color compression method based on hierarchical segmentation. First, this method performs global color segmentation. Second, the high-dimensional mean shift clustering algorithm [17] is used to cluster all points according to color. Finally, local segmentation is carried out in the geometric space to ensure that the resulting image blocks have color consistency (this feature is very useful for effective intra-frame prediction). Meanwhile, a skeleton-based motion estimation method proposed in Ref. [18] is used to derive the kinematic parameters of the human body for offline generation. The aim is to cluster the point cloud data into point sets related to the skeleton in order to obtain more accurate motion estimation and thus improve compression performance.

Rigid-body motion estimation is based on the Iterative Closest Point (ICP) algorithm [19], and it is also used for a similar purpose [20]. The motion-compensation residuals are stored in a residual map, which is generated by projecting the residuals onto a regular 2D grid embedded in a cylindrical coordinate system defined around the skeleton, and then they are compressed using a video codec. The advantage of using a skeleton is that it can be used to interact with objects in a virtual environment or analyze the motion of objects. However, since 3D point cloud data often contains noise, the correlation between points is not sufficient to support unique skeleton registration in 3D space. Because inaccurate motion compensation can lead to 2D residuals, it seriously affects the compression process and is not conducive to effective video coding.

Inspired by the geometrically driven method proposed in Ref. [21] and the compression-clustering-based technique proposed in Ref. [22], the method proposed in Ref. [23] combines geometric features and color features and clusters the point cloud into a set of planar patches. Then, a dedicated compression scheme is designed based on Gaussian Mixture Models (GMMs). The principle is to replace the clustered points with each planar patch using GMM approximation, and obtain multiple LODs by modifying the configuration of the model.

2.1.3 Point cloud compression methods based on graph transform

Graph-based transform methods extend traditional signal or image transform-based compression methods to point cloud data. The difficulty lies in the lack of topological information in point cloud representation. Wiman et al. [24] proposed a wavelet-based LiDAR-like (WAvelet Lidar Zipper, WALZ) point cloud compression framework. It uses wavelet transform to transform each coordinate component (x, y, z) into a one-dimensional vector. In this way, highly correlated and continuously recorded points will generate small and similar detail coefficients, which can be effectively compressed. Gu Shuai et al. [25] proposed a graph transform method. First, it is assumed that the point cloud is decomposed by an octree, and each occupied leaf node can be interpreted as a representative voxel. Then, the implicit topological structure of the voxel grid is used to connect adjacent points. Finally, a graph transform equivalent to the Karhunen-Loeve transform (KLT) [26] is adopted to decorrelate the geometric signals. However, its transformation performance highly depends on the topological information of the generated graph. In addition, since images may vary greatly between each frame, this method is not suitable for inter-frame encoding.

In Ref. [27], the introduced Region-Adaptive Hierarchical Transform (RAHT) aims to improve color compression performance. This method uses a hierarchical octree transform, grouping voxels into the lower levels of the octree to predict the colors of nodes in subsequent levels. When the prediction weights are consistent, this process can be simplified to the Haar transform. The main advantage of this method is its low computational complexity, making it suitable for real-time transmission scenarios. The Graph Fourier Transform (GFT) [28] was proposed based on their previous work [15]. This method takes advantage of the spatial correlation and temporal redundancy of color and geometric sequences. The method in Ref. [25] constructs a graph from the initial 3D point cloud with the help of octree decomposition and obtains the spectral representation of geometric and color signals using the GFT defined by the graph Laplacian operator.

The Gaussian Process Transform (GPT) proposed in Ref. [29] describes a method of modeling the statistics of point colors by assuming that point colors are samples of a stationary Gaussian process defined over an infinite 3D space. The covariance function of the Gaussian process is used to derive the covariance matrix of the colors, and then the Karhunen-Loeve Transform (KLT) of the color signal is obtained.

Shao et al. [30] proposed an optimized graph transform scheme for attribute compression. This method uses k-d tree partitioning to generate a graph, optimizes the Laplacian sparsity of the graph transform with the help of an offline training phase, and also takes into account the quantization process based on the RDO (Rate-Distortion Optimization) Lagrangian. Experimental results show that there are significant improvements in both transformation efficiency and R-D (Rate-Distortion) performance for static point clouds. However, in most cases, static point clouds require the creation of topological information, so the computational complexity of this type of method is relatively high. In the case of dynamic 3D point clouds, the complexity problem is particularly crucial, as additional 3D motion estimation or compensation procedures increase the computational burden. These methods cannot achieve the compression performance provided by the 2D-based V-PCC (Versatile Point Cloud Compression) [31] method, which makes full use of the efficiency of the HEVC (High-Efficiency Video Coding) [32] video codec in a practical way. However, the motion estimation stage carried out in 3D space is much more accurate than 2D video-based methods. It will be interesting to explore in future research how to leverage the accuracy of 3D motion estimation and the efficiency of 2D video-based technologies.

2.1.4 Point cloud compression methods based on learning

With the rapid development of deep learning techniques, point cloud geometric compression methods based on learning have emerged. Currently, most of the network architectures used in learning-based point cloud compression methods evolve from learning-based 2D image compression [33] network architectures. They use 3D convolution instead of 2D convolution in the meta-architecture and then process 3D voxelized point cloud data. Quach et al. [34], Xu et al. [35], and Guarda et al. [36] attempt to compress and decompress 3D occupancy model data of voxelized point cloud geometry using an autoencoder network based on 3D Convolution Neural Network (CNN). They optimize the decompression and reconstruction of occupied voxelized point cloud data modeled as a classification problem through Weight Binary Cross Entropy (WBCE) loss. Although these algorithms achieve good performance in point cloud geometric compression, they all have the problem of excessively high computational complexity in both the time and space dimensions.

This is because they rely on volumetric models (such as processing in blocks) without considering the sparsity of point clouds. In contrast, Yan et al. [37], Huang et al. [38], and Wen et al. [39] adopt point-based compression methods instead of the aforementioned voxelized point cloud representation. For example, Huang et al. [38] designed a hierarchical autoencoder with multi-scale loss for the reconstruction of fine-grained point cloud data. Wen et al. [39] proposed an adaptive octree compression network for large-scale point cloud representation. In these point-based methods, although redundant calculations can be reduced, since the correlations between voxels are not fully utilized, compared with MPEG G-PCC [40], there are only minor optimization improvements.

2.2 Three-dimensional point cloud sequence data coding and decoding

At present, the principle of point cloud sequence data coding and decoding technology based on 2D projection compression is to convert 3D point clouds into 2D videos through projection or mapping. Then, existing video coding technologies are utilized for compression.

In the method proposed by Ochotta et al. [41], the point cloud is first segmented into a set of patches and then projected onto a regular grid (referred to as point clusters). Subsequently, each cluster is parameterized as a height field on the corresponding base plane. Similar to the method in Ref. [42], using an atlas of height fields, heuristic segmentation, and merging the segmented point sets into patches, the height field is projected onto the base plane to obtain a 2D image. The 2D image is then compressed and decompressed using the wavelet-transform-based shape-adaptive image coding and decoding method proposed in Ref. [43]. Finally, the bit rate and quality are evaluated according to the actual network bandwidth, and optimal bit allocation is arranged in the rate – The compression method based on 2D projection proposed in Ref. [44] first projects a 2D grid pattern composed of horizontal and vertical lines onto a 3D model according to a set of viewpoints with the help of physical projection. Then, for each viewpoint, a set of 3D curves is generated, and these 3D curves are used as coding units. Finally, the point positions are prediction-coded using a spanning tree to achieve compression.

Several other 2D projection compression methods also adopt the multi-view representation method. The principle of the point cloud compression method based on multi-view representation is to first represent the point cloud as a set of depth images, then project them into multiple views corresponding to different viewing angles, and finally encode and decode the multi-views through image or video coding methods to achieve the compression of 3D point cloud data. The methods of Mekuria et al. [20], Merkle et al. [45], and Lien et al. [23] are multi-view video coding algorithms that utilize the statistical correlations between time and views within a prediction framework, distortion optimization step.

Ainala et al. [46] proposed a method combining octree and projection. The residuals obtained from planar projection are used to construct an enhancement layer within the framework of coarse octree coding. Through principal component analysis (PCA) of each cell of the octree, the points in the enhancement layer are projected onto a 2D plane. Then, predictive coding is performed on geometric and color attributes. In addition, to improve compression performance and reduce computational complexity, Ainala et al. [46] processed the generated prediction residuals using entropy coding techniques with the help of learning-based techniques. Compared with the JPEG-based point cloud data compression method proposed in Ref. [20], the compression ratio of the method proposed by Ainala et al. [46] has been significantly improved. Combining octree-based geometric compression and image-coding-based color compression has become an emerging research direction in recent years.

Reference [47] proposed a patch-based dynamic point cloud data compression standard, Video-Point Cloud Compression (V-PCC). The main idea of V-PCC is to utilize existing video codecs to compress the geometric and texture information of dynamic point clouds. V-PCC decomposes the input point cloud into a set of patches. First, with the help of local plane approximation, a normal vector is associated with each point. Then, within a given search radius, the initial clustering of points is determined by grouping adjacent points with similar directions (relative to a given threshold). The clustering is iteratively refined by updating adjacent points with similar normal vectors within a certain range to obtain smooth clustering boundaries and reduce related projection distortion. Since the points clustered in a cluster have similar normal vectors, a reference direction can be associated with each patch, which corresponds to the average of the individual normal vectors. Once the point cloud is divided into patches, this set of patches is vertically projected onto a 2D plane corresponding to the reference normal vector. The projection plane is discretized according to a regular sampling grid of a given resolution to obtain a pixelated representation of the projected image. Both depth and attribute (color) information are retained in the final projected image. Then, the 2D orthogonally projected patches are assembled into global images to generate geometric and attribute images. These patches are filled into the image in descending order of size. Once a patch is filled, the pixels are marked as "occupied". The bounding box information of the 2D patches is used to avoid intersections between patches while minimizing the blank space between patches in the global image. This filling method can result in a smoother image with stronger spatio-temporal consistency, making it more suitable for video coding.

2.3 Adaptive bit-rate coding and decoding technology

Currently, adaptive bit-rate coding and decoding technology are mainly applied in video coding and decoding, and there is no work yet applied to point cloud compression.

Recently, methods proposed by academic and industrial researchers [47–50] have gone beyond the fixed two-price ladder approach. All of these solutions rely on the coding statistics collected by performing a large number of encodings or using a selective number of encodings with pre-trained machine learning models to predict the bit-rate ladder.

Most of the methods introduced recently are based on an optimization framework for each title. In the work of [47, 48], they run multiple trial encodings at different quantization levels to obtain the Rate-Quantization (RQ) curves of each movie at different resolutions and different bit rates. Then, this information is used to construct the Pareto-optimal frontier of the RQ curve (commonly referred to as the convex hull in the adaptive streaming literature) using the scaled PSNR [47] and the scaled VMAF [48] so as to obtain the optimal parameters for a defined bit-rate range.

The method proposed in Ref. [51] measures the actual usage of millions of video clips to create a probability distribution of available bandwidth and viewport size. Compared with existing technologies, these probability distributions provide an optimized process that ensures video quality preservation while reducing the required bit rate. The solution of Bitmovin et al. [52] includes the calculation of encoding complexity. First, this method performs complexity analysis on each incoming video, and then various measurements are processed by a machine-trained model. Finally, the encoding profile is adjusted to match the content. An improved method is proposed in Ref. [53], which uses trial encodings to collect encoding statistics at low resolutions and utilizes them within a probabilistic framework to accelerate encoding decisions at high resolutions.

Recent work has proposed optimization methods based on each scene, with the aim of maximizing quality or minimizing the bit rate of each encoded representation in the video-on-demand scenario. This method builds a quantized convex hull by encoding sequences at a set of spatial resolutions. In addition, Netflix [49] updated their dynamic optimization method by establishing a bit-rate ladder through complexity analysis and further adjusted the predefined encoding parameters. Another interesting method is the one proposed by Brightcove [54, 55], which takes both quality constraints and bit-rate network statistics into account. The quality metric used in this example is the Structural Similarity Index Measure (SSIM), and the bit rate constraints are based on a probabilistic model. Another recent method is the iSize solution [56], which uses pre-encoding in a deep learning framework to determine the optimal set of encoding parameters and resolution at the block level.

While all of the above-mentioned solutions are important and contribute to the enhancement of video services, a direct and detailed comparison is not possible. Firstly, they are proprietary, which means they are designed to meet different business requirements and their full details may not be publicly available [57]. Additionally, some methods use different metrics in their bit-rate adaptation processes, and these metrics are not always shared. One aspect of improving these methods is to reduce their complexity, as most of the previously mentioned solutions rely on a large number of encodings (in many cases, a substantial amount).

3. Point cloud compression algorithms for vehicle recognition

3.1 Octree voxel-based point cloud compression method

This technique predominantly utilizes the closest neighboring point to the octree voxel's central point as a replacement for the central point itself. This substitution process serves to streamline the point cloud dataset, thereby achieving the objective of data compression. A significant benefit of this approach is its capacity to maintain the integrity of the original point cloud structure.

For a comprehensive point cloud dataset, let $P = \{P_1, P_2, \dots, P_n\}$ denote the total number of points within the dataset. Initially, determine the maximum and minimum values of the dataset along the three-dimensional coordinate axes. Subsequently, based on these values, construct a minimum bounding box, as described in Eq. (1):

$$(M)\,s = m_{i-1}m_{i-2}, \ldots, m_1m_0 \tag{1}$$

The minimum bounding box is linked to the root node of the octree. Afterward, this minimum bounding box is partitioned. Post-partitioning, each resulting rectangle is associated with a child node of the octree, and these child nodes are systematically encoded in octal format. The encoding scheme for the nodes is detailed in Eq. (2):

$$(M)_{10} = m_{i-1}8^{i-1} + m_{i-2}8^{i-2} + \ldots + m_18^1 + m_08^0 \tag{2}$$

The partition path of octree nodes is denoted from m_0 to m_{i-1}. The conversion relationship between the octal-encoded nodes and decimal encoding is presented in Eq. (2). The octree partition level is represented by i, and m_i is octal encoding, $m_0, m_{i-1} \in [0, 7]$.

Calculate the position of any point $P(x, y, z)$ in the octree in the point cloud data set, and mark the index values of its sub-nodes as e, f, and g.

$$\begin{cases} e = [(x - x_{min})/a] \\ f = [(y - y_{min})/b] \\ g = [(z - z_{min})/c] \end{cases} \tag{3}$$

The coordinate $(X_{min}, Y_{min}, Z_{min})$ represents the minimum vertex coordinate of the smallest bounding rectangle on the three-dimensional coordinate axis. Eq. (3) can be converted into binary representation as follows:

$$\begin{cases} e = a_{i-1}2^{i-1} + a_{i-2}2^{i-2} + \ldots + a_12^1 + a_02^0 \\ f = b_{i-1}2^{i-1} + b_{i-2}2^{i-2} + \ldots + b_12^1 + b_02^0 \\ g = c_{i-1}2^{i-1} + c_{i-2}2^{i-2} + \ldots + c_12^1 + c_02^0 \end{cases} \tag{4}$$

In the above equation, $(a_0, a_1, \ldots, a_{i-1}) \in \{0, 1\}$.

The cessation criterion for the octree partitioning process is predicated upon the voxel dimensions. In the event that the voxel size does not align with the predefined threshold, the recursive subdivision of the bounding box will persist; on the contrary, should the voxel size fall below the stipulated threshold, the partitioning will abruptly terminate. Following this determination, the positional indices of the octree voxels, as well as the centroid coordinates of these voxels, are meticulously computed and ascertained. Ultimately, the Euclidean distance metric is leveraged to pinpoint the point within the octree voxel that is in closest proximity to the centroid, and this proximate point is then designated to represent all other points within the voxel, thereby effectuating the compression of the point cloud dataset.

3.2 Point cloud boundary information extraction

Ascertain the status of a point as a boundary point by examining the angular relationships of the vectors resulting from the projections of the point in question and its neighboring points onto the tangent plane. A point is classified as a boundary point if there exists a substantial angular disparity between at least one pair of projection vectors, which is notably greater than the angles observed between other vector pairs. The principal steps of the algorithm are delineated as follows:

The process of plane fitting is executed for a given sampling point and its surrounding points. Let us designate the point under consideration as point P. Let us denote the identified point as point P, and the collection of its adjacent points as set $\mathrm{P_k} = \{\mathrm{P_0}, \mathrm{P_1}, \ldots, \mathrm{P_{k-1}}\}$.. Then, the least-squares fitting algorithm is applied.

The general form of the plane equation is presented as follows:

$$A * x + B * y + C * z + D = 0 \; (C \neq 0) \tag{5}$$

$$Z = -\frac{A}{C} * x - \frac{B}{C} * y - \frac{D}{C} = 0 \tag{6}$$

$$a_0 = -\frac{A}{C}, a_1 = -\frac{B}{C}, a_2 = -\frac{D}{C} \tag{7}$$

Substitute Eq. (7) into Eq. (6) to derive the following formula:

$$Z = a_0 * x + a_1 * y + a_2 \tag{8}$$

For a series of points n $(n \geq 3)$, $(x_i, y_i, z_i).i = 0, 1, \ldots, n-1$. To fit the plane equation with the point n, that is, making Eq. (9) the smallest, it should meet $a\frac{\partial S}{\partial a_k} = 0, k = 0, 1, 2.$

$$S = \sum_{i=0}^{n-1} (a_0 * x + a_1 * y + a_2 - z)^2 \tag{9}$$

$$\begin{cases} \sum 2(a_0 * x_i + a_1 * y_i + a_2 - z_i) * x_i = 0 \\ \sum 2(a_0 * x_i + a_1 * y_i + a_2 - z_i) * y_i = 0 \\ \sum 2(a_0 * x_i + a_1 * y_i + a_2 - z_i) = 0 \end{cases} \tag{10}$$

If Eq. (10) is changed to matrix form, it is:

$$\begin{vmatrix} \sum x_i^2 & \sum x_i y_i & \sum x_i \\ \sum x_i y_i & \sum y_i^2 & \sum y_i \\ \sum x_i & \sum y_i & n \end{vmatrix} \begin{bmatrix} a_0 \\ a_1 \\ a_2 \end{bmatrix} = \begin{bmatrix} \sum x_i z_i \\ \sum y_i z_i \\ \sum z_i \end{bmatrix} \tag{11}$$

To address the task, we will follow these steps:

1. Solve the System of Equations:

   Begin by solving the system of equations derived from the aforementioned formula to determine the parameters a_0, a_1, a_2. of the plane.

2. Substitute Parameters into the Plane Equation:

   Once the parameters are obtained, substitute them into the plane equation to fully define the plane.

3. Calculate Projections:

   With the plane equation in hand, calculate the projections of the sampling point and its neighboring points onto the plane.

4. Define the Sampling Point:

   Let the sampling point be denoted as p.

5. Fit a Vertical Line Through the Point:

   Fit a vertical line through point p that is perpendicular to the plane.

6. Determine the Foot of the Perpendicular:

   Let the foot of the perpendicular from point $p'(x', y', z')$ to the plane be denoted as pp'.

7. Relate the Line to the Normal Vector:

   The line pp' is parallel to the normal vector $\vec{n}$ of the plane.

8. Derive the Normal Vector from the Plane Equation:

The normal vector $\vec{n}$ of the plane can be derived directly from the plane equation $\vec{n} = (A, B, C)$.

The parametric equation of p is as follows:

$$\begin{aligned} x &= x_i - At \\ y &= y_i - Bt \\ z &= z_i - Ct \end{aligned} \tag{12}$$

Bring the point (x', y', z') into the plane equation to obtain t:

$$t = \frac{Ax_i + By_i + Cz_i + D}{A_2 + B_2 + C_2} \tag{13}$$

Then substitute t into Eq. (12) to obtain the coordinates of the projection point p'.

The next procedure involves identifying and removing the edge points from the collection of projected points. After determining the coordinates of these projected points, consider the set of projected points for the k nearest neighbors of a given point to be denoted as $p'_k = \{p'_0, p'_1, \dots, p'_{k-1}\}$. Subsequently, create vectors $p'p'_0, p'p'_1, \dots, p'p'_{k-1}$, totaling k vectors, which are used in this segment to capture the concept of boundaries. To ascertain if a point p is an edge point, compare the angles enclosed by k the vectors.

The process of calculating the vector angle is as follows:

1. Starting with a particular point p' as the origin, consider all points in the set of points p'_k as destinations, and construct k vectors. Choose any one of these vectors as the reference vector. For computational simplicity, this paper selects as $p'p'_0$ the reference vector. Initially, calculate the cross-product of $p'p'_0$ with the normal vector of the plane to obtain vector 1

2. Obtain the angles α_i and β_i between vector $p'p'_i$ and vector 1, respectively, with $p'p'_i (i = 1, 2, \dots, k)$ in turn. If $\beta_i > 90^\circ$, then $\alpha_i = 360 - \alpha_i$.

3. Arrange the α_i obtained in (2) in ascending order, and calculate the angle γ_i

4. between the two vectors $p'p'_0$ and $p'p'_i(i = 1, 2, \ldots, k)$, according to α_i, as shown in Eq. (14):

$$\gamma_i = \begin{cases} \alpha_i(i = 1) \\ \alpha_i - \alpha_{i-1}(i = 2, 3, \ldots, k - 2) \\ 360 - \alpha_i(i = k - 1) \end{cases} \tag{14}$$

5. Compare the value γ_i calculated in (3), take the maximum value, and compare it with the set angle threshold. If it is greater than the set threshold, then the corresponding point p'_i is the boundary point and is added to the boundary point set.

3.3 Key feature point extraction

By examining the normal vectors across various regions of the point cloud, it is observed that areas with gradual changes in normal vectors tend to be relatively flat, whereas areas with rapid changes in normal vectors are characterized by steep and undulating terrain. The characteristic degree at a point p_i, which signifies the trend of normal vector variation, is defined as the arithmetic mean of the angles between the normal vector at point p_i and the normal vectors at its k nearest neighbor points. The formula is shown as Eq. (16), where p_j is the nearest neighbor point of point p_i and the angle between the normal vector of point p_i and p_j is θ_{ij}:

$$f_i = \frac{1}{k}\sum_{j=1}^{k} \theta ij \tag{15}$$

To encapsulate the main idea, the more significant the trend of change, the more distinct the local features become. Consequently, we can harness this characteristic to pinpoint the essential feature points within the point cloud. Select a more appropriate threshold ε_1, filter out the relatively flat regions in the point cloud, and take $f_1 > \varepsilon_1$ as the condition of the reserved point. If it is satisfied, it is reserved. For any point p_m that meets the reserved condition, if there is:

$$f(p_m) = \max[f(p_{m1}), f(p_{m2}), \ldots, f(p_{mk})] \tag{16}$$

Then p_m will be the characteristic point, and the characteristic degree of the k nearest neighbor of point p_m is $f(p_{m1}), f(p_{m2}), \ldots, f(p_{mk})$, p is point cloud data, and obtain the data set $p_i = \{p_{i1}, p_{i2}, \ldots, p_{im}\}$ of p by feature extraction.

By combining the nearest-neighbor points of the previously determined center points, the boundaries of the point cloud, and the already extracted feature points, the process of point cloud compression is achieved.

3.4 Image mapping point cloud vehicle recognition

In this instance, YOLO-v5 is utilized as the core network for image recognition tasks. The pixel coordinates within the bounding box of the detected image targets are matched with the corresponding point cloud data, effectively refining the search area for the point cloud targets. Following this, the vehicle point cloud data is sifted using the bounding box of the vehicle targets as determined by logistic regression.

For the point cloud vehicle recognition, an image-mapped algorithm is implemented. The process starts with the deployment of the YOLO algorithm to perform an initial identification of vehicles present in the image. Then, a mapping is created between the image and the compressed point cloud data, facilitated by the calibration matrix and the projection relationships between different data acquisition devices. This mapping allows for the filtering of target points. Finally, the K-means clustering algorithm is fine-tuned to precisely identify the vehicle targets within the point cloud environment.

The specific recognition techniques employed in this scenario are detailed below:

1. Recognition of vehicles within images using the YOLO algorithm.

2. Establishing a correspondence between point cloud data and image data.

3. Detection of targets within the point cloud using the K-means clustering algorithm.

In this case, YOLO-v5 is leveraged as the primary network for image recognition. The pixel coordinates within the bounding box of the image target are correlated with the point cloud data, which helps to reduce the search scope for the point cloud target. Following this, the vehicle point cloud is refined by using the bounding box of the vehicle target that is generated by logistic regression.

3.5 Experiments

To more effectively assess the algorithm for vehicle recognition from image convolution and its correlation to point cloud object recognition, as presented in this paper, the image data from the KITTI dataset can be used to train the convolutional network. The process begins by converting the image data from the KITTI dataset from PNG to VOC format. Then, this data is fed into the YOLO network model. Lastly, the calibration file provided in the KITTI dataset is used to calibrate the converted data.

The experiment is grounded in the YOLO-v5 algorithm. The findings reveal that the YOLO-v5 algorithm has a notably high detection rate for vehicle targets within images. Additionally, it holds significant advantages over the majority of recognition

Figure 1.
KITTI image vehicle recognition results.

Figure 2.
Mapping of point cloud data in image.

networks that process point cloud data as input. As shown in **Figure 1**, precise vehicle target detection is achievable by identifying vehicles in image data across a variety of settings.

In the experiment, a selective point cloud is utilized, which corresponds exclusively to the image and excludes other elements. By utilizing the conversion relationship between the lidar and the camera, as provided in the KITTI dataset, the alignment of the point cloud to the image is accomplished. The outcome of transforming the point cloud into image coordinates through coordinate transformation is depicted in **Figure 2**.

By employing the rotation matrix that relates the point cloud data to the image, the point cloud data can be sifted using the pre-existing bounding box from the YOLO algorithm. In this context, the blue points indicate the point cloud that has been aligned with the vehicle pixel points within the image, as illustrated in **Figure 3**.

Sometimes, targets can overlap. During recognition, the bounding boxes might repeatedly pick up frames, potentially causing the loss of the point cloud target's position. Moreover, during recognition, the bounding boxes might also capture other interferences, such as vegetation, steps, and ground noise. Consequently, after the point cloud filtering, the K-means clustering algorithm is applied to refine the target. This process helps to preserve the vehicle data, minimize the impact of other environmental noises, and improve the positional accuracy of the point cloud vehicle recognition.

Figure 3.
Image pixel mapping to point cloud data.

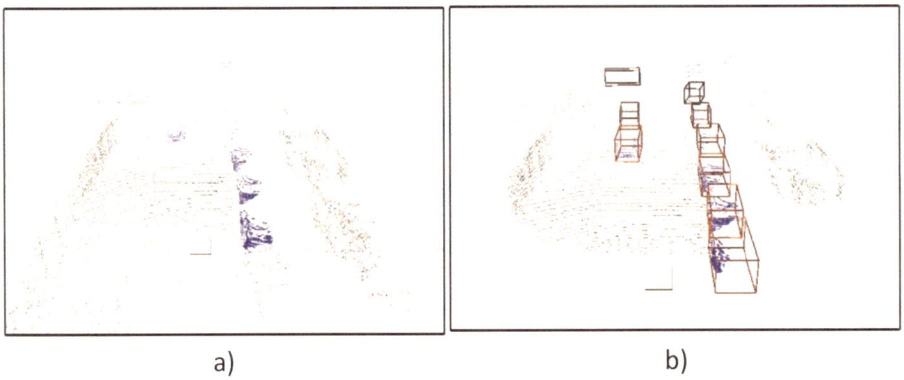

Figure 4.
Vehicle point cloud clustering and bounding box calculation (a) Vehicle Point Cloud Clustering (b) Calculate Bounding Box.

As depicted in **Figure 4**, by performing iterative calculations with the K-means clustering algorithm, the vehicle clustering results, which are mapped from image pixels to point cloud data, are obtained. In addition, the mean vector for each of the K vehicle clusters is also derived.

4. Conclusions

At present, in the field of point cloud data compression research, most work is focused on compressing small-scale scene point clouds. However, the environments captured by vehicle-mounted LiDAR systems are typically large-scale open-air settings such as streets and power line corridors. In these scenarios, the volume of point cloud data is significantly greater than that of isolated scenes. Additionally, the scenes captured by vehicles contain a variety of ground objects like trees, vehicles, roads, and buildings, leading to a high degree of scene complexity.

Although traditional coding and decoding technologies are capable of performing the coding and decoding processes on vehicle-mounted LiDAR point cloud data, numerous issues still persist, mainly manifested in the following aspects:

1. The currently mainstream coding and decoding methods are unable to effectively decompress and reconstruct accurate point cloud data. Additionally, the computational complexity of the models is high.

2. Given the complex nature of vehicle-mounted LiDAR point cloud sequence data, the existing point cloud sequence coding and decoding methods cannot effectively compress each object in an appropriate manner. Moreover, the process of converting three-dimensional data into a two-dimensional format for video compression gives rise to problems such as high bit-rate, low quality, and low timeliness.

3. The current point cloud compression models based on end-to-end optimization lack the ability to arbitrarily adjust the bit rate. This renders the encoding technology inapplicable in practical situations.

At present, deeply integrating artificial intelligence methods with LiDAR point cloud data coding and decoding, exploring learning-based LiDAR point cloud data compression algorithms, proposing an autonomous learning mechanism for point cloud sequence data coding and decoding, and studying the working mechanism of high-efficiency point cloud sequence data compression with adaptive bit rate can further enrich and develop existing basic theories, such as deep learning, information theory, and signal processing. This endeavor holds significant theoretical significance and practical application value, providing theoretical support and technical methods for the autonomous learning of LiDAR point cloud sequence coding and decoding.

Author details

Yanjun Zhang* and Xiaoyu Sun
Zhongshan Polytechnic, Information and Engineering School, Zhongshan, China

*Address all correspondence to: yj.zhang@zspt.edu.cn

IntechOpen

© 2025 The Author(s). Licensee IntechOpen. This chapter is distributed under the terms of the Creative Commons Attribution License (http://creativecommons.org/licenses/by/4.0), which permits unrestricted use, distribution, and reproduction in any medium, provided the original work is properly cited. (cc) BY

References

[1] Caihong D, Yao Z. Point cloud neighborhood search algorithm based on octree encoding. Computer Engineering and Design. 2018;**10**:3107-3122

[2] Garcia DC, de Queiroz RL. Intra-frame context-based octree coding for point-cloud geometry. In: 2018 25th IEEE International Conference on Image Processing (ICIP). Athens, Greece: IEEE; 2018. pp. 1807-1811

[3] Sayood K. Introduction to data compression. San Francisco, USA: Morgan Kaufmann Publishers Inc.; 2017

[4] Anand B, Barsaiyan V, Senapati M, et al. Real time lidar point cloud compression and transmission for intelligent transportation system. In: 2019 IEEE 89th Vehicular Technology Conference (VTC2019-Spring). Kuala Lumpur, Malaysia: IEEE; 2019. pp. 1-5

[5] Garcia DC, de Queiroz RL. Context-based octree coding for point-cloud video. In: 2017 IEEE International Conference on Image Processing (ICIP). Beijing, China: IEEE; 2017. pp. 1412-1416

[6] Ohl S. Tele-immersion concepts. IEEE Transactions on Visualization and Computer Graphics. Los Alamitos, USA: IEEE Computer Society. 2017;**24**(10): 2827-2842

[7] Kathariya B, Li L, Li Z, et al. Scalable point cloud geometry coding with binary tree embedded quadtree. In: 2018 IEEE International Conference on Multimedia and Expo (ICME). San Diego, USA: IEEE; 2018. pp. 1-6

[8] Yan R, Zongju P, Xin C, et al. Fast partitioning of FVC intra - coding units combining random forest. Journal of Image and Graphics. 2019;**24**(5):724-733

[9] Huang Y, Peng J, Kuo CCJ, et al. A generic scheme for progressive point cloud coding. IEEE Transactions on Visualization and Computer Graphics. Los Alamitos, USA: IEEE Computer Society. 2008;**14**(2):440-453

[10] Yuxue F. Research on simplification and compression techniques for 3D models with multiple attributes. Shandong University. 2017. [Doctoral thesis]

[11] Kitago M, Gopi M. Efficient and prioritized point subsampling for CSRBF compression. In: SPBG. Boston, Massachusetts, USA: Symposium on Point Based Graphics; 2006. pp. 121-128

[12] Fengtao W, Fengyi L. Application of fusion kernel function in improved radial basis function surrogate model. Computer Engineering and Applications. 2019:7:58-65

[13] Golla T, Klein R. Real-time point cloud compression. In: 2015 IEEE/RSJ International Conference on Intelligent Robots and Systems (IROS). Hamburg, Germany: IEEE; 2015. pp. 5087-5092

[14] De Queiroz RL, Chou PA. Motion-compensated compression of dynamic voxelized point clouds. IEEE Transactions on Image Processing. 2017; **26**(8):3886-3895

[15] Thanou D, Chou PA, Frossard P. Graph-based motion estimation and compensation for dynamic 3D point cloud compression. In: 2015 IEEE International Conference on Image Processing (ICIP). Quebec, Canada: IEEE; 2015. pp. 3235-3239

[16] Zhang K, Zhu W, Xu Y. Hierarchical segmentation based point cloud attribute compression. In: 2018 IEEE International

Conference on Acoustics, Speech and Signal Processing (ICASSP). QLD, Australia: IEEE; 2018. pp. 3131-3135

[17] Liwei C, Yanfang Q, Haifeng Z, et al. Endmember bundle extraction based on mean shift clustering. Applied Science and Technology. 2020;**47**(1):21-30

[18] Shujie L, Zhu H, Wang L, et al. Dual auto - encoder network for human skeletal motion data optimization. Journal of Image and Graphics. 2022;**4**: 1277-1289

[19] Danni G, Shaoxia P, Modi H, et al. Comparison of 3D facial scan registration methods applied to edentulous implant restoration design. Journal of Peking University (Health Sciences). 2021;**53**(1):83

[20] Mekuria R, Blom K, Cesar P. Design, implementation, and evaluation of a point cloud codec for tele-immersive video. IEEE Transactions on Circuits and Systems for Video Technology. 2016; **27**(4):828-842

[21] Taubin G, Rossignac J. Geometric compression through topological surgery. ACM Transactions on Graphics (TOG). 1998;**17**(2):84-115

[22] Morell V, Orts S, Cazorla M, et al. Geometric 3D point cloud compression. Pattern Recognition Letters. 2014;**50**: 55-62

[23] Viejo D, Cazorla M. A robust and fast method for 6DoF motion estimation from generalized 3D data. Autonomous Robots. 2014;**36**(4):295-308

[24] Wiman H, Qin Y. Fast compression and access of lidar point clouds using wavelets. In: 2009 Joint Urban Remote Sensing Event. ShangHai, China: IEEE; 2009. pp. 1-6

[25] Shuai G, Huanqiang Z, Jing C, et al. 3D point cloud compression using improved graphical transformation. Signal Processing. 2019;**35**(1):32-38

[26] Zhang C, Florêncio D. Analyzing the optimality of predictive transform coding using graph-based models. IEEE Signal Processing Letters. 2012;**20**(1): 106-109

[27] De Queiroz RL, Chou PA. Compression of 3D point clouds using a region-adaptive hierarchical transform. IEEE Transactions on Image Processing. 2016;**25**(8):3947-3956

[28] Thanou D, Chou PA, Frossard P. Graph-based compression of dynamic 3D point cloud sequences. IEEE Transactions on Image Processing. 2016; **25**(4):1765-1778

[29] De Queiroz RL, Chou PA. Transform coding for point clouds using a Gaussian process model. IEEE Transactions on Image Processing. 2017;**26**(7):3507-3517

[30] Shao Y, Zhang Z, Li Z, et al. Attribute compression of 3D point clouds using Laplacian sparsity optimized graph transform. In: 2017 IEEE Visual Communications and Image Processing (VCIP). SuZhou, China: IEEE; 2017. pp. 1-4

[31] Schwarz S, Preda M, Baroncini V, et al. Emerging MPEG standards for point cloud compression. IEEE Journal on Emerging and Selected Topics in Circuits and Systems. 2018;**9**(1):133-148

[32] Martínez-Rach MO, Migallón H, López-Granado O, et al. Performance overview of the latest video coding proposals: HEVC, JEM and VVC. Journal of Imaging. 2021;**7**(2):39

[33] Minnen D, Ballé J, Toderici G. Joint autoregressive and hierarchical priors for

learned image compression. In: The 32nd Conference on Neural Information Processing Systems. Montreal, Canada: NIPS; 2018

[34] Quach M, Valenzise G, Dufaux F. Learning convolutional transforms for Lossy pointcloud geometry compression. In: 2019 IEEE International Conference on Image Processing (ICIP). Taibei, China: IEEE; 2019. pp. 4320-4324

[35] Jiacheng X, Fang Z, Gao Y, Ma S, Jin Y, Zhou H, et al. Point AE-DCGAN: A deep learning model for 3D point cloud Lossy geometry compression. In: 2021 Data Compression Conference (DCC). Utah, USA: IEEE; 2021. p. 379. DOI: 10.1109/DCC50243.2021.00085

[36] Guarda AFR, Rodrigues NMM, Pereira F. Point cloud coding: Adopting a deep learning-based approach. In: 2019 Picture Coding Symposium (PCS). San Francisco, USA: IEEE; 2019. pp. 1-5

[37] Yan W, Liu S, Li TH, et al. Deep autoencoder-based Lossy geometry compression for point clouds. Computer Vision and Pattern Recognition, arXivLabs. 2019

[38] Huang T, Liu Y. 3d point cloud geometry compression on deeplearning. In: Proceedings of the 27th ACM International Conference on Multimedia. Nice, France. 2019. pp. 890-898

[39] Wen X, Wang X, Hou J, et al. Lossy geometry compression of 3d point cloud data via an adaptive octree-guided network. In: 2020 IEEE International Conference on Multimedia and Expo (ICME). London, UK: IEEE; 2020. pp. 1-6

[40] Mammou K, Chou PA, Flynn D, et al. G-pcc codec description v2. ISO/IEC JTC1/SC29/WG11 N18189. 2019

[41] Ochotta T, Saupe D. Compression of Point-Based 3D Models by Shape-Adaptive Wavelet Coding of Multi-Height Fields. 2004

[42] Labussière M, Laconte J, Pomerleau F. Geometry preserving sampling method based on spectral decomposition for large-scale environments. Frontiers in Robotics and AI. 2020;7:572-578

[43] Shu Z, Yao Q, Xie Z, et al. Arbitrarily shape object-based video coding technology by finite Ridgelet transform. In: 2018 IEEE International Conference on Information and Automation (ICIA). Wuyishan, China: IEEE; 2018. pp. 1520-1523

[44] Daribo I, Furukawa R, Sagawa R, et al. Efficient rate-distortion compression of dynamic point cloud for grid-pattern-based 3D scanning systems. 3D Research. 2012;**3**(1):2

[45] Merkle P, Smolic A, Muller K, et al. Multi-view video plus depth representation and coding. In: 2007 IEEE International Conference on Image Processing. Vol. 1. San Antonio, TX, USA: IEEE; 2007. pp. I-201-I-204

[46] Ainala K, Mekuria RN, Khathariya B, et al. An improved enhancement layer for octree based point cloud compression with plane projection approximation. In: Applications of Digital Image Processing XXXIX. International Society for Optics and Photonics. Vol. 9971. 2016. p. 99710R

[47] De Cock J, Li Z, Manohara M, et al. Complexity-based consistent-quality encoding in the cloud. In: 2016 IEEE International Conference on Image Processing (ICIP). Phoenix, USA: IEEE; 2016. pp. 1484-1488

[48] Afonso M, Moorthy A, Guo L, et al. Improving our video encodes for legacy devices. The Netflix Tech Blog. 2020

[49] Katsavounidis I. Dynamic Optimizer–A Perceptual Video Encoding Optimization Framework. The Netflix Tech Blog; 2018

[50] Timmerer C, Hellwagner H. Mipso: Multi-period per-scene optimization for Http adaptive streaming. In: 2020 IEEE International Conference on Multimedia and Expo. (ICME). London, UK: IEEE; 2020. pp. 1-6

[51] Chen C, Lin YC, Benting S, et al. Optimized transcoding for large scale adaptive streaming using playback statistics. In: 2018 25th IEEE International Conference on Image Processing (ICIP). Athens, Greece: IEEE; 2018. pp. 3269-3273

[52] Bitmovin. White Paper: Per Title Encoding. 2018. Available from: https: bitmovin.com/whitepapers/bitmovin-per-title.pdf

[53] Goswami K, Hariharan B, Ramachandran P, et al. Adaptive multi-resolution encoding for ABR streaming. In: 2018 25th IEEE International Conference on Image Processing (ICIP). Vol. 2018. Athens, Greece: IEEE. pp. 1008-1012

[54] Reznik YA, Lillevold KO, Jagannath A, et al. Optimal design of encoding profiles for ABR streaming. In: Proceedings of the 23rd Packet Video Workshop. 2018. pp. 43-47

[55] Reznik YA, Li X, Lillevold KO, et al. Optimal multi-codec adaptive bitrate streaming. In: 2019 IEEE International Conference on Multimedia & Expo Workshops (ICMEW). Vol. 2019. Shanghai, China: IEEE. pp. 348-353

[56] Bourtsoulatze E, Chadha A, Fadeev I, et al. Deep video precoding. IEEE Transactions on Circuits and Systems for Video Technology. 2019; **30**(12):4913-4928

[57] Bentaleb A, Taani B, Begen AC, et al. A survey on bitrate adaptation schemes for streaming media over HTTP. IEEE Communications Surveys & Tutorials. 2018;**21**(1):562-585

Chapter 8

Artificial Intelligence-Driven Electric Vehicle Battery Lifetime Diagnostics

Jingyuan Zhao and Andrew F. Burke

Abstract

Ensuring the reliability, safety, and efficiency of electric vehicles (EVs) necessitates precise diagnostics of battery life, as the degradation of batteries directly influences both performance and sustainability. The transformative role of artificial intelligence (AI) in advancing EV battery diagnostics is explored herein, with an emphasis placed on the complexities of predicting and managing battery health. Initially, we provide an overview of the challenges associated with battery lifetime diagnostics, such as issues with accuracy, generalization, and model training. The following sections delve into advanced AI methodologies that enhance diagnostic capabilities. These methods include extensive time-series AI, which improves predictive accuracy; end-to-end AI, which simplifies system complexity; multi-model AI, which ensures generalization across varied operating conditions; and adaptable AI strategies for dynamic environments. In addition, we explore the use of federated learning for decentralized, privacy-preserving diagnostics and discuss the role of automated machine learning in streamlining the development of AI-based models. By integrating these sophisticated AI techniques, we present a comprehensive roadmap for the future of AI-driven battery prognostics and health management. This roadmap underscores the critical importance of accuracy, scalability, and sustainability in fostering advancement. Our interdisciplinary framework offers valuable insights that can accelerate the electrification of transportation and advance the evolution of energy storage systems, tackling key challenges at the intersection of battery technology and AI.

Keywords: battery, artificial intelligence, electric vehicle, machine learning, health

1. Introduction

Research into battery lifetime diagnostics is crucial, as it directly impacts the reliability, safety, and economic viability of electric vehicles (EVs), particularly in addressing the challenges posed by complex degradation mechanisms and operational conditions [1]. With the global surge in EV adoption, there is a heightened focus on understanding and managing battery degradation—a complex phenomenon driven by material properties (such as the degradation mechanisms of different anode and cathode materials), operational conditions, and environmental factors that directly impact battery capacity and power fade throughout its lifecycle [2]. Effective diagnostics are crucial for mitigating performance loss, extending battery life, and promoting the sustainability of EV technologies. Battery deterioration in EVs arises from intricate physical, chemical, and electrochemical activities. Key processes include the formation of a solid electrolyte interphase (SEI), the deposition of lithium on electrode surfaces, and alterations in electrode microstructures [3]. The complexity of these processes varies significantly across different battery types, such as lithium iron phosphate (LFP), nickel manganese cobalt oxide, and lithium nickel cobalt aluminum oxide (NCA) [4, 5]. It is crucial to deploy sophisticated diagnostic tools capable of accurately monitoring the cell health and projecting its remaining useful life (RUL) under a variety of operational conditions.

Traditionally, the evaluation of EV battery health relied on established physical models, which stem from core principles of electrochemistry and materials science [6–8]. While these models provide deep insights, their application is often constrained by limited scalability and adaptability in diverse, real-world environments. To address these shortcomings, the field has increasingly turned to artificial intelligence (AI) and machine learning [9–13]. Contemporary diagnostic strategies are increasingly incorporating sophisticated AI techniques. Deep learning models such as convolutional neural networks (CNNs), recurrent neural networks (RNNs), and physics-informed neural networks are now at the forefront of battery diagnostics [14–16]. These technologies excel in processing and analyzing complex data arrays, providing precise assessments of battery health even when data availability is limited. This shift not only improves the accuracy of diagnostics but also adapts more fluidly to the diverse conditions encountered in practical EV usage. Developing robust diagnostic tools also requires high-quality data, which forms the cornerstone of modeling and prediction. However, standardization of data collection and harmonization of diagnostic methods are significant challenges in the field, and building a community of data centers with standardized practices and flexible sharing options may offer great opportunities for AI in battery lifetime diagnostics [17–19]. Moreover, diagnostic advancements are supported by new sensing technologies such as impedance spectroscopy, differential voltage analysis, and real-time thermal monitoring, enriching the data pools available for AI models.

Integrating these diagnostics into battery management system (BMS) not only aids in predictive maintenance but also aligns with energy efficiency and sustainability objectives. By facilitating accurate battery health monitoring and management, diagnostics are critical to reducing waste, optimizing resource use, and supporting the circular economy through battery reuse and recycling, especially in the context of Industry 4.0 [20]. As EV demand grows, developing accurate, scalable, and efficient diagnostic technologies remains fundamental to the progress of transportation electrification.

2. Challenges of EV battery lifetime diagnostics

EV battery lifetime diagnostics encounter numerous obstacles (**Figure 1**). These challenges stem from the complex nature of battery systems, the variety of operating

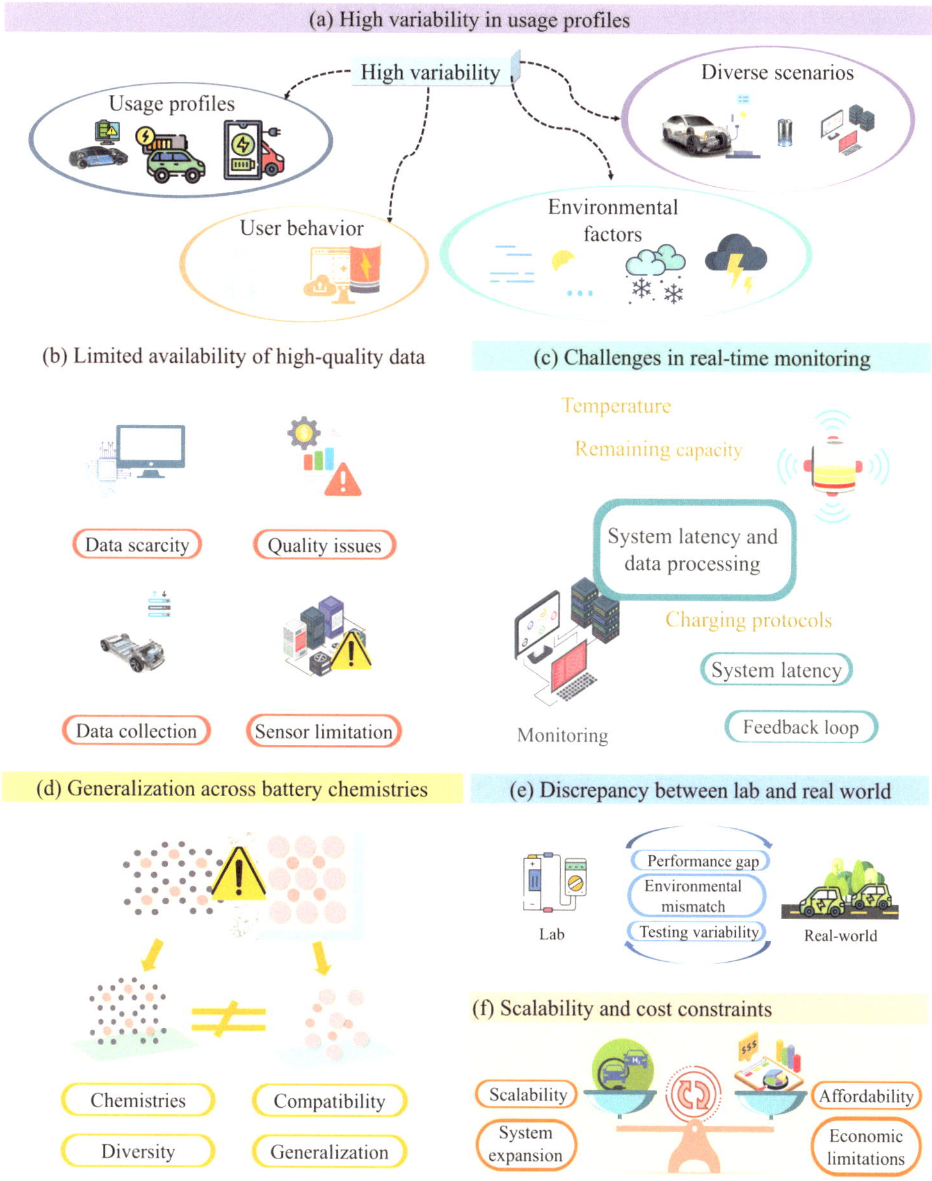

Figure 1.
Challenges in battery health and lifetime diagnostics for EVs.

conditions, and the evolving processes of battery aging. These obstacles are generally grouped into three categories: technical, methodological, and practical. To overcome these issues, a multidisciplinary strategy is essential, requiring combined knowledge in electrochemistry, materials science, data science, and AI. Developing sophisticated diagnostic frameworks is crucial. Such frameworks must strike a balance between accuracy, computational efficiency, and applicability to maintain the reliability, safety, and performance of EV batteries throughout their service life [21, 22]. This section will address these six aspects in detail.

2.1 High variability in usage profiles

EV batteries operate under highly varied and dynamic conditions shaped by user behaviors, including charging frequency, driving patterns, ambient temperatures, and load demands. Frequent fast charging accelerates lithium plating and intensifies localized thermal gradients, while irregular discharge cycles create uneven mechanical and chemical stress across active materials [23, 24]. These factors contribute to non-uniform cell degradation, complicating the development of predictive diagnostic models.

A significant challenge lies in translating laboratory test data, typically gathered under controlled and static conditions, to real-world scenarios characterized by stochastic and non-repeating patterns [25]. Laboratory tests often use constant-current or constant-power protocols, which fail to capture transient events such as regenerative braking, high-power bursts, or prolonged idle periods. These real-world dynamics generate localized reactions and degradation pathways not adequately represented by standardized laboratory datasets. Environmental variability further compounds the issue. Extreme temperatures influence ionic conductivity and reaction rates, introducing spatial and temporal inconsistencies in degradation [26]. In cold climates, lithium-ion diffusion slows, increasing resistance, while in hot climates, accelerated electrolyte decomposition promotes faster capacity fade. Diagnostic models must integrate thermal-electrochemical coupling to account for these effects and accurately predict performance across a broad range of operating conditions. The lack of high-resolution data from real-world fleet operations poses another obstacle. Data sparsity limits the ability to train robust models, particularly for emerging chemistries and applications. To address this, techniques such as physics-informed generative models and transfer learning are increasingly employed to augment datasets and enhance predictive accuracy [27, 28]. Hybrid frameworks combining machine learning with rule-based methods also allow the integration of domain knowledge, improving the ability to manage variability in usage conditions.

To generalize across diverse usage profiles, diagnostic models must combine robustness to variability with computational efficiency for implementation in embedded systems. Advances in reinforcement learning and adaptive algorithms offer promising solutions, enabling real-time model calibration using operational data. However, scaling these approaches to account for the vast array of conditions encountered in real-world EV applications remains a critical area for future research.

2.2 Limited availability of high-quality data

Developing robust diagnostic models necessitates extensive datasets capturing a broad range of battery chemistries, operational conditions, and aging trajectories. However, obtaining such datasets remains a significant challenge. Battery manufacturers and EV companies often treat data as proprietary assets, restricting access to detailed performance records, such as precise charge-discharge cycles, temperature profiles, and health metrics. This lack of transparency hampers research efforts to develop universally applicable models. Real-world datasets, when available, present additional challenges [29]. These datasets are often incomplete, with irregular sampling intervals, missing critical variables, or limited coverage of edge cases such as extreme fast-charging cycles or long-term idle states. Noise introduced by sensor inaccuracies or communication delays in BMS further complicates data usability. The lack of standardization in data acquisition protocols across manufacturers exacerbates the problem, resulting in datasets that are difficult to integrate or compare [17]. Another critical limitation is the scarcity of degradation data across the entire battery lifecycle. Laboratory testing typically emphasizes specific aging conditions, such as constant-current cycling or fixed-temperature environments, which fail to reflect the nuanced degradation patterns observed in field conditions. Long-term data from batteries used in real-world applications, particularly under diverse operating conditions, is rare but essential for accurately training and validating predictive models.

To address these challenges, researchers are leveraging advanced data augmentation and synthetic data generation techniques. Physics-informed generative models, such as those combining electrochemical modeling with machine learning, have shown promise in creating realistic degradation scenarios from limited datasets [30, 31]. Transfer learning methods are models trained on one dataset that are able to realize knowledge migration under different chemical or operational conditions by fine-tuning adaptation to another dataset, thus providing an effective solution to the problems of training data scarcity and cross-domain adaptation [32]. In addition, federated learning frameworks are emerging as a solution to access distributed datasets from multiple stakeholders without requiring direct data sharing, preserving privacy while enriching model diversity [33]. Despite these advances, achieving comprehensive datasets that capture the full spectrum of real-world conditions and aging pathways remains an unmet need. Efforts to standardize data collection protocols and establish collaborative industry-academia data-sharing agreements could significantly accelerate progress in developing more accurate and generalizable battery diagnostic models.

2.3 Challenges in real-time monitoring

Real-time diagnostics in EVs require achieving a delicate balance between computational efficiency and diagnostic accuracy to provide actionable insights without compromising system performance. Onboard BMS, constrained by limited processing power, memory, and energy budgets, must accommodate models capable of rapid inference while maintaining high predictive precision. Advanced diagnostic methods, such as deep learning and physical information modeling, excel in dealing with the complex dynamics of multi-physics fields and multi-scale battery systems, but the reliance of these methods on large-scale computation and high-dimensional data processing poses a significant challenge for their deployment in resource-limited environments [34]. For instance, CNNs and recurrent architectures such as long short-term memory networks, commonly used for state of health (SOH) estimation, require significant computational resources for training and inference. Techniques such as quantization, pruning, and model compression are increasingly employed to reduce model size and energy consumption [35, 36]. Furthermore, hybrid approaches, integrating neural networks with reduced-order physical models, provide a promising path to retain critical electrochemical insights while lowering computational overhead [37]. Latency is another critical concern in real-time monitoring, particularly during transient events such as fast charging or high-power discharges. Models must generate predictions within milliseconds to support dynamic real-time applications, where delays can critically impact system performance and user experience. Recent advancements in efficient attention mechanisms, such as sparse-attention and flash-attention, have emerged as transformative solutions for handling high-frequency data streams. These techniques not only reduce computational complexity but also optimize memory usage and enhance numerical stability, enabling models to process longer sequences and maintain high throughput. By balancing efficiency with accuracy, these attention mechanisms have demonstrated their suitability for real-time implementations across diverse domains, such as autonomous systems, financial forecasting, and large-scale language models, where rapid decision-making is essential [28, 38].

Real-time monitoring of battery health poses several challenges that complicate the deployment of reliable diagnostics. Key issues include data sparsity and the presence of noise in sensor readings, often due to environmental factors, measurement errors, or hardware limitations. These conditions require the development of diagnostic systems that can function effectively despite suboptimal data quality. To address these challenges, sophisticated signal processing techniques are commonly used. For instance, Kalman filtering is adept at estimating battery states by integrating noisy and intermittent data, while denoising autoencoders use neural networks to restore corrupted signals to their original state. These methods are crucial for deriving actionable insights from flawed data. In addition, real-time adaptive learning frameworks are vital for ensuring the accuracy of diagnostics as battery conditions change. Techniques such as online learning allow models to assimilate new data incrementally, avoiding the need for comprehensive retraining [39]. Reinforcement learning, on the other hand, enables models to adjust dynamically, optimizing decisions through trial-and-error interactions with the battery system [40]. Integrating domain knowledge, such as electrochemical models, with AI-based diagnostics further bolsters the reliability of these systems. Hybrid approaches that merge physics-based insights with data-driven algorithms enhance the interpretability and precision of predictions, especially in data-limited scenarios. Confronting these challenges is essential for maintaining the robustness and effectiveness of real-time monitoring systems. Successfully navigating these issues not only improves the reliability and safety of battery systems but also supports the implementation of advanced technologies across various operational contexts.

2.4 Generalization across battery chemistries

Generalizing diagnostic models across battery chemistries, such as LFP, nickel cobalt manganese (NCM), and NCA, is a complex challenge due to the distinct electrochemical behaviors, aging mechanisms, and operational characteristics of each type. For example, LFP batteries exhibit relatively stable voltage profiles and are less prone to thermal runaway, but their aging is dominated by capacity fade due to phase boundary evolution [41]. In contrast, NCM and NCA batteries, known for their higher energy densities, face more pronounced capacity fade from electrolyte decomposition and transition metal dissolution, as well as heightened risks of thermal instability [42].

Key technical hurdles arise from variations in degradation pathways and sensitivities to operating conditions. For instance, NCM batteries are particularly affected by high-temperature operation, which accelerates the growth of the SEI and transition metal migration, while LFP batteries are more tolerant to higher temperatures but suffer from performance degradation at low temperatures due to lithium-ion diffusivity limitations [26]. These differing characteristics demand diagnostic models that can capture both shared and chemistry-specific features to provide accurate predictions across all battery types. Another challenge lies in manufacturing variations. Differences in electrode formulations, electrolyte additives, and production processes significantly impact cell performance and aging [43, 44]. Even within the same chemistry, cells from different manufacturers may exhibit distinct degradation trajectories, making it difficult to develop a universal model. This issue is further exacerbated by limited labeled data for emerging chemistries or proprietary battery designs, which restrict the training and validation of diagnostic algorithms.

Transfer learning and multi-task learning frameworks offer potential solutions by leveraging shared representations across chemistries while allowing for task-specific fine-tuning. For instance, pre-trained models on large datasets of LFP cells can be adapted to NCA cells with minimal retraining, provided key cross-chemistry features are effectively captured [45]. In addition, hybrid diagnostic approaches that integrate physics-based modeling with data-driven techniques can enhance generalizability by embedding fundamental electrochemical principles applicable to all chemistries. Advanced feature extraction techniques, such as those leveraging attention mechanisms in transformer architectures [46], enable the identification of critical patterns unique to specific chemistries while maintaining a global understanding of shared dynamics. Moreover, federated learning approaches, where models are collaboratively trained across distributed datasets without direct data sharing, can address the challenge of manufacturer-specific variations while preserving data privacy. Ensuring reliable diagnostics across chemistries also requires addressing scalability. Models must not only generalize to new chemistries but also adapt to variations in battery configurations, such as differences in cell formats (pouch, cylindrical, prismatic) or pack-level integration. Developing scalable, modular frameworks that accommodate such variations will be critical for creating truly universal battery diagnostic systems.

2.5 Discrepancy between lab and real-world conditions

Models validated in controlled laboratory settings often fail to capture the complexities and variabilities of real-world conditions [9, 47]. Laboratory protocols, such as constant-current or constant-power cycling, are designed to assess battery performance and degradation but do not mimic irregular usage patterns driven by behaviors such as regenerative braking, partial state-of-charge operations, and variable depths of discharge. These real-world conditions can lead to transient phenomena such as voltage hysteresis, uneven current distribution, and intermittent thermal gradients that are seldom reproduced in laboratory environments [48]. Furthermore, environmental factors significantly affect diagnostics in practical settings. Extreme temperatures can drastically alter battery performance and lifespan; for example, cold conditions may decrease ion mobility and increase resistance, leading to lithium plating, while high temperatures might hasten electrolyte decomposition and SEI growth [26]. In addition, temperature gradients within battery packs, often resulting from inconsistent cooling, introduce variations in cell aging that laboratory conditions typically do not replicate. Mechanical stresses, including vibrations and shocks during vehicle operations, introduce another layer of complexity. These forces can compromise the mechanical structure of electrodes, lead to electrolyte leakage, or create micro-cracks in separators, impacting that laboratory tests rarely emulate but which significantly influence battery safety and longevity.

Bridging the gap between laboratory and field conditions necessitates innovative approaches that integrate the variability of real-world environments into model development. Physics-informed machine learning is particularly promising, blending fundamental electrochemical principles with adaptations to patterns observed in actual usage [49]. Moreover, hybrid models that combine laboratory data with real-time telemetry from the field allow for more effective generalization across diverse conditions. For instance, data from onboard sensors can recalibrate models initially trained in labs, enhancing their relevance and adaptability [50]. Initiatives to better replicate real-world conditions through advanced testing protocols are becoming more common [51]. Cycling profiles that incorporate dynamic loads, fluctuating temperatures, and partial states of charge are now used to simulate more realistic battery operation conditions. Integrating tools such as impedance spectroscopy or thermal imaging into practical applications also facilitates ongoing validation and refinement of models.

Developing accurate diagnostics for real-world applications requires solutions that can effectively manage operational complexities without compromising the precision achieved in laboratory models. Strategies such as transfer learning enable models trained on controlled datasets to adapt to diverse field conditions. Federated learning, which allows models to learn collaboratively across decentralized datasets without directly sharing sensitive information, can help incorporate variability from different operating environments. Adaptive algorithms, capable of real-time updates based on new data, further improve the flexibility and reliability of diagnostics under changing and unpredictable conditions. These approaches collectively advance the applicability of diagnostic models in practical settings.

2.6 Scalability and cost constraints

Scaling diagnostic solutions for the growing EV market demands a balance between computational efficiency, accuracy, and cost-effectiveness. Diagnostic algorithms must be lightweight enough to run on embedded processors in BMS while maintaining performance comparable to more resource-intensive cloud-based models. Achieving this requires innovative algorithmic designs and hardware-software co-optimization.

One key approach involves model compression techniques, such as pruning, quantization, and knowledge distillation, which reduces the size and complexity of diagnostic models while preserving critical functionality. Pruning eliminates redundant connections in neural networks, while quantization reduces the precision of model parameters to lower bit-width formats [52, 53]. Knowledge distillation allows a smaller, efficient model to learn from a larger, high-capacity model, enabling high-performance diagnostics on low-power hardware. In addition, edge computing is increasingly utilized to process data locally within the vehicle, minimizing the need for continuous data transmission to the cloud. This reduces operational costs and latency while addressing privacy concerns. Specialized hardware accelerators, such as application-specific integrated circuits and field-programmable gate arrays, are being adopted to further enhance computational efficiency for real-time diagnostics [54]. These accelerators can optimize specific diagnostic tasks, such as SOH estimation or thermal runaway prediction, by leveraging parallel processing and energy-efficient designs. Scalability also requires reducing reliance on extensive training data, which can be both costly and time-consuming to collect. Few-shot learning and transfer learning methods allow diagnostic models to adapt quickly to new battery chemistries or operational profiles with minimal labeled data. For instance, a model trained on one type of battery, such as NCM, can be fine-tuned for LFP batteries, significantly reducing development costs.

From an operational perspective, implementing hierarchical diagnostic frameworks can improve scalability. These frameworks divide diagnostic tasks into tiers, with simpler computations performed on local processors and more complex analyses delegated to centralized servers during maintenance cycles. This modular approach optimizes resource allocation, ensuring cost efficiency at scale. Lastly, advanced simulation environments, such as digital twins, enable the development and validation of scalable diagnostic solutions before deployment. By mirroring the behavior of real-world systems, digital twins allow researchers to test algorithms under diverse conditions, reducing the need for extensive field trials [55]. This approach accelerates innovation while controlling costs, ensuring diagnostic solutions remain practical for widespread adoption in the rapidly expanding EV market.

3. Cutting-edge AI methods for battery lifetime diagnostics

Recent advancements in AI have revolutionized the field of battery lifetime diagnostics, enabling more precise and efficient tools to assess and predict battery health. As batteries operate under diverse and dynamic conditions, their degradation processes often exhibit nonlinear and complex patterns. Traditional models, while valuable, struggle to adapt to these complexities and scale effectively to real-world applications. Cutting-edge AI approaches address these challenges by leveraging advanced architectures designed to handle large datasets, dynamic inputs, and intricate relationships within battery systems. The following subsections explore some pivotal AI methods (**Figure 2**) that are driving innovation in battery lifetime diagnostics. Each of these approaches represents a critical step forward in optimizing battery diagnostics and supporting the broader goals of electrification and sustainability.

3.1 Large time-series AI for improving prediction accuracy

Managing the vast amount of time-series data generated by modern batteries is essential to understanding their health and predicting their performance. Large time-series AI models are particularly well-suited to this challenge, as they are capable of processing and analyzing data patterns over extended periods. Batteries undergo continuous cycles of charging and discharging, and the data recorded over these cycles includes critical variables such as voltage, current, and temperature. These variables reflect the cell operational behavior and aging process, making accurate prognostics both complex and vital for optimizing battery use. Large time-series AI models use deep learning architectures that can efficiently handle the high dimensionality and temporal dependencies in this data. By analyzing patterns across multiple cycles, these models can detect gradual degradation trends and subtle anomalies that indicate potential failure or performance reduction. For instance, transformer-based architectures or hybrid models combining convolutional and recurrent layers are particularly adept at capturing both short-term fluctuations and long-term degradation trends, providing more precise forecasts for metrics such as SOH and RUL [56, 57]. A key advantage of these models is their ability to perform predictive diagnostics, identifying signs of deterioration early enough to allow for timely interventions. By recognizing the early warning signals embedded in the data, these models help reduce the risk of unexpected battery failures and extend operational lifespans. This is especially critical in applications such as EVs and energy storage systems, where reliability and performance are paramount. Furthermore, large time-series AI models offer scalability, enabling their application across diverse battery chemistries and usage scenarios [28]. Their ability to generalize insights from vast datasets means that they can adapt to different battery technologies, from lithium-ion to newer chemistries, while still providing accurate and actionable predictions. As battery systems grow more

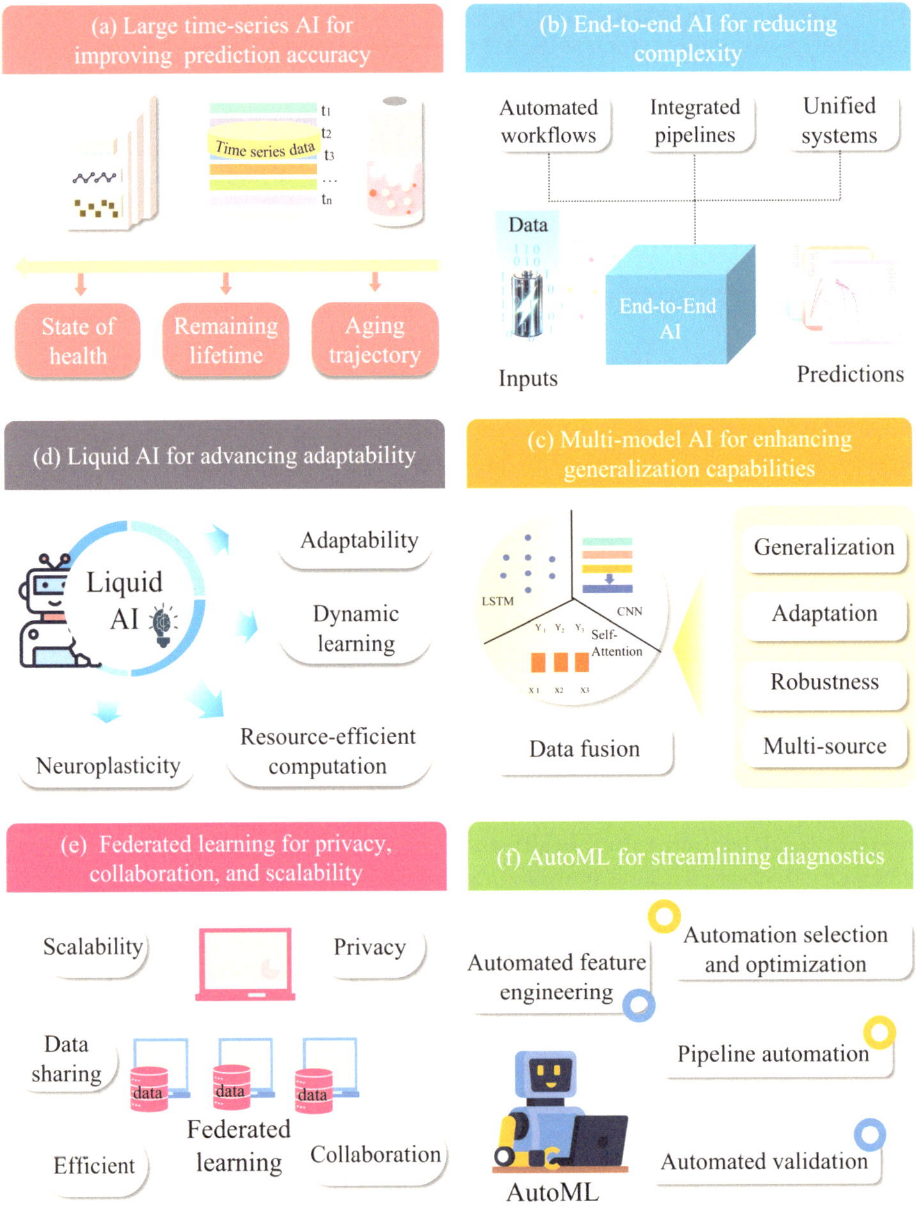

Figure 2.
Revolutionizing battery prognostics and health management through large AI models.

complex, large time-series AI will be a cornerstone technology for improving prognostics. These models allow for smarter, data-driven decisions, ultimately enhancing battery safety, reliability, and efficiency in real-world applications.

3.2 End-to-End AI for reducing complexity

End-to-End AI introduces a seamless approach to battery prognostics and management by integrating all stages of data analysis into a unified framework. Rather than relying on traditional methods that require multiple steps—such as manual feature extraction, data preprocessing, and separate model training—end-to-end AI systems handle these processes holistically. This enables a streamlined workflow from raw data input directly to actionable predictions, making it highly effective in managing the complex nature of battery systems. In battery applications, these AI models can directly process raw time-series data, such as voltage, current, and temperature readings, from the battery itself [58, 59]. By automatically learning relevant patterns and features, End-to-End AI eliminates the need for human intervention in feature engineering, thereby reducing the risk of missing critical insights. This automated approach allows the models to better capture trends and anomalies that might indicate changes in the SOH or predict the RUL of the battery.

The ability of End-to-End AI systems to continuously learn and update based on new data is crucial for real-time monitoring and predictive maintenance. In environments such as EVs, where operational conditions can vary significantly, End-to-End AI models can analyze live data streams and provide immediate insights into battery health [60, 61]. These real-time predictions enable operators to make timely decisions, such as adjusting charging protocols to extend battery lifespan or detecting early signs of malfunction, which reduces the risk of failure. Moreover, End-to-End AI is highly adaptable, capable of integrating various data sources for a more comprehensive view of battery systems. By combining operational data with environmental factors—such as temperature or humidity—it enhances the accuracy of its predictions, offering a full-spectrum analysis of how different conditions affect battery performance over time. Another major advantage of End-to-End AI lies in its scalability. Whether applied to small-scale consumer electronics, EVs, or large-scale energy storage systems, these models can be adapted to suit different battery types and use cases with minimal adjustments. This flexibility allows for precise and efficient management of battery health across various applications.

3.3 Multi-model AI for enhancing generalization capabilities

Multi-model AI refers to the integration of multiple AI models that operate collaboratively to improve the performance and robustness of predictions. In battery prognostics and management, this approach leverages the strengths of various models—such as neural networks, decision trees, and physics-based models—to handle the inherent complexity and variability of battery systems. The advantage of multi-model AI lies in its ability to address different aspects of battery behavior. For example, CNNs may excel in feature extraction from time-series data, while RNNs or transformers capture long-term dependencies in battery aging trends. Combining these models can leverage their complementary strengths, enabling the creation of hybrid architectures that extract local features efficiently while simultaneously modeling temporal dependencies, resulting in improved accuracy and robustness for battery health prediction and aging trend analysis [62, 63]. Furthermore, by combining these approaches with physical information models, multi-model AI systems can effectively integrate domain knowledge and observational data to provide more accurate and resilient predictions even under different operating conditions [31, 37]. This approach reduces the risk of single-model overfitting and enhances generalization across different battery chemistries, applications, and environments. In practical applications, multi-model AI can be particularly valuable for real-time monitoring and diagnostics. For example, by fusing the outputs of different models and combining physical information with a deep learning framework, it allows for continuous tuning of the parameters, which significantly improves the predictive accuracy and applicability of the model under multi-dynamic operating conditions [64]. This capability ensures more reliable SOH predictions, early fault detection, and optimized maintenance schedules. Looking ahead, the integration of multi-model AI into BMS will play a critical role in advancing the reliability and efficiency of next-generation energy storage technologies. As AI models continue to evolve, multi-model frameworks will likely become the standard for achieving comprehensive, high-fidelity diagnostics, enabling more effective management of battery systems across industries.

3.4 Liquid AI for advancing adaptability

Liquid AI represents a significant advancement in neural network technology [65], offering high adaptability to real-time data and variable conditions, which is crucial for effective battery management. This AI model is unique in its capability to dynamically modify its operations to meet changing demands without extensive retraining. Its applicability in battery management is demonstrated by its ability to smoothly adjust to fluctuations in battery use, environmental changes, and aging. For example, liquid neural networks have shown effectiveness in autonomous navigation by concentrating on pertinent features and disregarding irrelevant ones, allowing them to handle vastly different situations [66]. This attribute is beneficial in battery systems, where it helps dynamically accommodate operational changes and aging, ensuring precise predictions of SOH and dependable functionality. This adaptability contributes to consistent accuracy in forecasting SOH and potential degradation, thus improving the operational efficiency and lifespan of battery systems. Moreover, liquid AI is versatile, applicable across various battery types and uses. In EVs, it customizes management approaches based on individual driving patterns and charging habits, enhancing battery utilization and health over time. In the context of renewable energy storage, it adjusts dynamically to the patterns of energy production and consumption, maintaining stability and efficiency in power grid operations. This flexibility underscores the potential of liquid AI to revolutionize battery management across different sectors, ensuring both enhanced performance and sustainability.

The potential for liquid AI to integrate into battery diagnostic and management frameworks is vast. Its ongoing learning and adaptive capabilities are poised to minimize maintenance costs and maximize the performance and safety of battery systems. With continued development, liquid AI is expected to become a fundamental technology in next-generation battery management solutions, offering scalable, efficient responses to the diverse needs of the energy sector.

3.5 Federated learning for decentralized, privacy-preserving applications

Federated learning offers a transformative approach to AI-based battery diagnostics by enabling collaborative model training across decentralized datasets while preserving data privacy [33, 67]. In traditional AI frameworks, raw data from various sources must be aggregated into a central repository for training, posing challenges related to data ownership, security, and regulatory compliance. Federated learning overcomes these limitations by keeping data localized on individual devices or systems, such as EVs or distributed energy storage units, while sharing only model updates. This ensures that proprietary or sensitive data remains secure and private, addressing one of the critical barriers to scaling AI-based solutions in the battery domain.

Federated learning offers significant benefits for large-scale deployments involving diverse battery chemistries, usage patterns, and operational conditions that produce varied datasets. This approach supports decentralized learning, allowing systems to contribute to model enhancement while maintaining data integrity [68]. As a result, it generates universally applicable models that provide reliable diagnostics and predictions across numerous applications. In addition, federated learning efficiently handles data from geographically scattered EVs by performing model training locally, which eliminates the need to directly share sensitive data [69]. This collaborative method not only enhances the accuracy of estimating battery SOH but also adapts to varying environmental conditions and driving behaviors, all while protecting the privacy of the data providers. The privacy-preserving nature of federated learning not only ensures data security but also aligns with regulatory requirements in industries where data sharing is restricted. Moreover, it promotes collaborative innovation among stakeholders, including manufacturers, fleet operators, and researchers, fostering the development of more effective and reliable BMS. As battery technologies and deployment scales continue to evolve, federated learning offers a scalable, secure, and efficient framework for advancing AI-driven diagnostics, making it a cornerstone for the next generation of battery prognostics and health management (PHM) solutions.

3.6 AutoML for streamlining AI-based diagnostics

Automated machine learning (AutoML) is transforming the development and deployment of AI models for battery PHM by streamlining time-consuming and complex tasks. AutoML reduces the need for manual intervention in critical stages such as model selection, hyperparameter tuning, and feature engineering [70]. This shift allows researchers and engineers to focus more on strategic AI applications rather than the complexities of model construction. The automation speeds up the development cycle and ensures that models are finely tuned to meet the specific needs of diverse battery datasets.

For battery PHM, where datasets often vary greatly in size, structure, and complexity due to different chemistries, operating conditions, and degradation behaviors, AutoML can be a highly valuable tool. It automatically determines the most effective model configurations for particular datasets, facilitating the creation of customized solutions that enhance both accuracy and efficiency. For example, AutoML efficiently selects the optimal neural network architecture or adjusts hyperparameters to improve SOH estimations or RUL predictions for specific battery types [71]. The benefits of AutoML go beyond mere time savings. By lowering the technical threshold required for deploying AI, AutoML is particularly beneficial for teams and organizations without in-depth AI expertise [72]. In addition, it enhances scalability, allowing for the easy adaptation of diagnostic models to a wide array of battery applications—from small personal devices to extensive EV fleets and large-scale energy storage systems.

As battery technologies and their applications evolve, the development and implementation of specialized, high-performance AI models become increasingly essential. AutoML facilitates these processes by streamlining model development while ensuring accuracy and robustness, making it an indispensable tool in advancing AI-driven battery PHM. This method ensures that AI benefits are broadly accessible and effective, supporting a wide array of battery applications in a dynamic market.

4. Conclusions and outlook

AI has transformed the diagnostics of EV battery lifetimes, enhancing the monitoring, prediction, and management of battery health. AI enables advanced analysis of degradation patterns and dynamic performance metrics, overcoming many limitations of traditional models to improve precision, scalability, and adaptability. ML effectively captures the complexities of battery behavior, supporting accurate SOH assessments and robust RUL predictions. These advances are key in improving BMS and optimizing performance across various conditions. As AI evolves, its role in diagnostics is poised to grow. Future research should focus on expanding datasets, particularly those reflecting real-world usage, to enhance model robustness and generalization. Integrating multimodal data streams, including thermal, electrical, and mechanical sensor information, will enhance diagnostic capabilities. Real-time AI systems will enable immediate diagnostics and adaptive control strategies, ensuring continuous optimal operation. Self-learning algorithms with lifelong learning capabilities will adapt alongside evolving technologies, maintaining accuracy without frequent retraining. Sustainability will play a critical role in AI-driven diagnostics. By extending lifespans, optimizing usage, and enabling second-life applications, AI will reduce the environmental impact of EV batteries, supporting resource efficiency and circular economy practices. With growing EV demand, AI solutions will scale production, enhance reliability, and promote sustainable growth in electrification. The convergence of AI and battery research promises significant breakthroughs, improving economic and operational viability and driving innovation in energy and transportation.

Author details

Jingyuan Zhao* and Andrew F. Burke
Institute of Transportation Studies, University of California Davis, CA, USA

*Address all correspondence to: jyzhao@ucdavis.edu

IntechOpen

© 2025 The Author(s). Licensee IntechOpen. This chapter is distributed under the terms of the Creative Commons Attribution License (http://creativecommons.org/licenses/by/4.0), which permits unrestricted use, distribution, and reproduction in any medium, provided the original work is properly cited. (cc) BY

References

[1] Hu X, Xu L, Lin X, et al. Battery lifetime prognostics. Joule. 2020;**4**(2):310-346. DOI: 10.1016/j.joule.2019.11.018

[2] Han X, Lu L, Zheng Y, et al. A review on the key issues of the lithium ion battery degradation among the whole life cycle. ETransportation. 2019;**1**:100005. DOI: 10.1016/j.etran.2019.100005

[3] Palacín MR, de Guibert A. Why do batteries fail? Science. 2016;**351**(6273):1253292. DOI: 10.1126/science.1253292

[4] Wang L, Qiu J, Wang X, et al. Insights for understanding multiscale degradation of $LiFePO_4$ cathodes. EScience. 2022;**2**(2):125-137. DOI: 10.1016/j.esci.2022.03.006

[5] Rinkel BLD, Vivek JP, Garcia-Araez N, et al. Two electrolyte decomposition pathways at nickel-rich cathode surfaces in lithium-ion batteries. Energy & Environmental Science. 2022;**15**(8):3416-3438. DOI: 10.1039/D1EE04053G

[6] Li J, Adewuyi K, Lotfi N, et al. A single particle model with chemical/mechanical degradation physics for lithium ion battery state of health (SOH) estimation. Applied Energy. 2018;**212**:1178-1190. DOI: 10.1016/j.apenergy.2018.01.011

[7] Bi Y, Yin Y, Choe SY. Online state of health and aging parameter estimation using a physics-based life model with a particle filter. Journal of Power Sources. 2020;**476**:228655. DOI: 10.1016/j.jpowsour.2020.228655

[8] Xiong R, Li L, Li Z, et al. An electrochemical model based degradation state identification method of lithium-ion battery for all-climate electric vehicles application. Applied Energy. 2018;**219**:264-275. DOI: 10.1016/j.apenergy.2018.03.053

[9] Zhao J, Qu X, Wu Y, et al. Artificial intelligence-driven real-world battery diagnostics. Energy and AI. 2024;**18**:100419. DOI: 10.1016/j.egyai.2024.100419

[10] Zhang Y, Tang Q, Zhang Y, et al. Identifying degradation patterns of lithium ion batteries from impedance spectroscopy using machine learning. Nature Communications. 2020;**11**(1):1706. DOI: 10.1038/s41467-020-15235-7

[11] Zhao J, Ling H, Liu J, et al. Machine learning for predicting battery capacity for electric vehicles. ETransportation. 2023;**15**:100214. DOI: 10.1016/j.etran.2022.100214

[12] Zhao J, Feng X, Pang Q, et al. Battery prognostics and health management from a machine learning perspective. Journal of Power Sources. 2023;**581**:233474. DOI: 10.1016/j.jpowsour.2023.233474

[13] Khaleghi S, Hosen MS, Van Mierlo J, et al. Towards machine-learning driven prognostics and health management of Li-ion batteries. A comprehensive review. Renewable and Sustainable Energy Reviews. 2024;**192**:114224. DOI: 10.1016/j.rser.2023.114224

[14] He N, Wang Q, Lu Z, et al. Early prediction of battery lifetime based on graphical features and convolutional neural networks. Applied Energy. 2024;**353**:122048. DOI: 10.1016/j.apenergy.2023.122048

[15] Lu J, Xiong R, Tian J, et al. Battery degradation prediction against uncertain future conditions with recurrent neural network enabled deep learning. Energy Storage Materials. 2022;**50**:139-151. DOI: 10.1016/j.ensm.2022.05.007

[16] Wang F, Zhai Z, Zhao Z, et al. Physics-informed neural network for lithium-ion battery degradation stable modeling and prognosis. Nature Communications. 2024;**15**(1):4332. DOI: 10.1038/s41467-024-48779-z

[17] Ward L, Babinec S, Dufek EJ, et al. Principles of the battery data genome. Joule. 2022;**6**(10):2253-2271. DOI: 10.1016/j.joule.2022.08.008

[18] Li D, Nan J, Burke AF, et al. Battery prognostics and health management: AI and big data. World Electric Vehicle Journal. 2024;**16**(1):10. DOI: 10.3390/wevj16010010

[19] Dos Reis G, Strange C, Yadav M, et al. Lithium-ion battery data and where to find it. Energy and AI. 2021;**5**:100081. DOI: 10.1016/j.egyai.2021.100081

[20] Zhao J, Burke AF. Battery prognostics and health management for electric vehicles under industry 4.0. Journal of Energy Chemistry. 2023;**84**:30-33. DOI: 10.1016/j.jechem.2023.04.042

[21] Li H. Practical evaluation of Li-ion batteries. Joule. 2019;**3**(4):911-914. DOI: 10.1016/j.joule.2019.03.028

[22] Zhao J, Feng X, Pang Q, et al. Battery safety: Machine learning-based prognostics. Progress in Energy and Combustion Science. 2024;**102**:101142. DOI: 10.1016/j.pecs.2023.101142

[23] Jangid MK, Mukhopadhyay A. Real-time monitoring of stress development during electrochemical cycling of electrode materials for Li-ion batteries: Overview and perspectives. Journal of Materials Chemistry A. 2019;**7**(41):23679-23726. DOI: 10.1039/C9TA06474E

[24] Zhao J, Lv Z, Li D, et al. Battery engineering safety technologies (BEST): Mechanisms, modes, metrics, modelling and mitigation. ETransportation. 2024;**22**:100364. DOI: 10.1016/j.etran.2024.100364

[25] Wang Z, Shi D, Zhao J, et al. Battery health diagnostics: Bridging the gap between academia and industry. eTransportation. 2024;**19**:100309. DOI: 10.1016/j.etran.2023.100309

[26] Hou J, Yang M, Wang D, et al. Fundamentals and challenges of lithium ion batteries at temperatures between− 40 and 60° C. Advanced Energy Materials. 2020;**10**(18):1904152. DOI: 10.1002/aenm.201904152

[27] Zhang S, Liu Z, Xu Y, et al. A physics-informed hybrid data-driven approach with generative electrode-level features for lithium-ion battery health prognostics. IEEE Transactions on Transportation Electrification. 2024;**11**:4857-4871. DOI: 10.1109/TTE.2024.3471626

[28] Zhao J, Wang Z. Specialized convolutional transformer networks for estimating battery health via transfer learning. Energy Storage Materials. 2024;**71**:103668. DOI: 10.1016/j.ensm.2024.103668

[29] Sulzer V, Mohtat P, Aitio A, et al. The challenge and opportunity of battery lifetime prediction from field data. Joule. 2021;**5**(8):1934-1955. DOI: 10.1016/j.joule.2021.06.005

[30] Li H, Zhang Z, Li T, et al. A review on physics-informed data-driven remaining useful life prediction:

Challenges and opportunities. Mechanical Systems and Signal Processing. 2024;**209**:111120. DOI: 10.1016/j.ymssp.2024.111120

[31] Zhang S, Liu Z, Xu Y, et al. A physics-informed hybrid multitask learning for lithium-ion battery full-life aging estimation at early lifetime. IEEE Transactions on Industrial Informatics. 2024;**21**:415-424. DOI: 10.1109/TII.2024.3452273

[32] Pan SJ, Yang Q. A survey on transfer learning. IEEE Transactions on Knowledge and Data Engineering. 2009;**22**(10):1345-1359. DOI: 10.1109/TKDE.2009.191

[33] Zhang C, Xie Y, Bai H, et al. A survey on federated learning. Knowledge-Based Systems. 2021;**216**:106775. DOI: 10.1016/j.knosys.2021.106775

[34] Zhao J, Han X, Ouyang M, et al. Specialized deep neural networks for battery health prognostics: Opportunities and challenges. Journal of Energy Chemistry. 2023;**87**:416-438. DOI: 10.1016/j.jechem.2023.08.047

[35] Ge Y, Ma J, Sun G. A structural pruning method for lithium-ion batteries remaining useful life prediction model with multi-head attention mechanism. Journal of Energy Storage. 2024;**86**:111396. DOI: 10.1016/j.est.2024.111396

[36] Choudhary T, Mishra V, Goswami A, et al. A comprehensive survey on model compression and acceleration. Artificial Intelligence Review. 2020;**53**:5113-5155. DOI: 10.1007/s10462-020-09816-7

[37] Nascimento RG, Corbetta M, Kulkarni CS, et al. Hybrid physics-informed neural networks for lithium-ion battery modeling and prognosis. Journal of Power Sources. 2021;**513**:230526. DOI: 10.1016/j.jpowsour.2021.230526

[38] Lv Z, Zhao J. Resource-efficient artificial intelligence for battery capacity estimation using convolutional flash attention fusion networks. eTransportation. 2025;**23**:100383. DOI: 10.1016/j.etran.2024.100383

[39] Shen S, Sadoughi M, Chen X, et al. A deep learning method for online capacity estimation of lithium-ion batteries. Journal of Energy Storage. 2019;**25**:100817. DOI: 10.1016/j.est.2019.100817

[40] Chun H, Han S. Maximizing the performance of a lithium-ion battery aging estimator using reinforcement learning. IEEE Transactions on Industrial Informatics. 2024;**20**:7019-7029. DOI: 10.1109/TII.2024.3353861

[41] Eddahech A, Briat O, Vinassa JM. Performance comparison of four lithium–ion battery technologies under calendar aging. Energy. 2015;**84**:542-550. DOI: 10.1016/j.energy.2015.03.019

[42] Kvasha A, Gutiérrez C, Osa U, et al. A comparative study of thermal runaway of commercial lithium ion cells. Energy. 2018;**159**:547-557. DOI: 10.1016/j.energy.2018.06.173

[43] Liu C, Lombardo T, Xu J, et al. An experimentally-validated 3D electrochemical model revealing electrode manufacturing parameters' effects on battery performance. Energy Storage Materials. 2023;**54**:156-163. DOI: 10.1016/j.ensm.2022.10.035

[44] Zhu P, Slater PR, Kendrick E. Insights into architecture, design and manufacture of electrodes for lithium-ion batteries. Materials & Design. 2022;**223**:111208. DOI: 10.1016/j.matdes.2022.111208

[45] Zhao J, Qu X, Han X, et al. Cross-material battery capacity estimation using hybrid-model fusion transfer learning. Journal of Power Sources. 2025;**625**:235674. DOI: 10.1016/j.jpowsour.2024.235674

[46] Zhao J, Han X, Wu Y, et al. Opportunities and challenges in transformer neural networks for battery state estimation: Charge, health, lifetime, and safety. Journal of Energy Chemistry. 2024;**102**:463-496. DOI: 10.1016/j.jechem.2024.11.011

[47] Qu X, Shi D, Zhao J, et al. Insights and reviews on battery lifetime prediction from research to practice. Journal of Energy Chemistry. 2024;**94**:716-739. DOI: 10.1016/j.jechem.2024.03.013

[48] Zhao J, Feng X, Tran MK, et al. Battery safety: Fault diagnosis from laboratory to real world. Journal of Power Sources. 2024;**598**:234111. DOI: 10.1016/j.jpowsour.2024.234111

[49] Navidi S, Thelen A, Li T, et al. Physics-informed machine learning for battery degradation diagnostics: A comparison of state-of-the-art methods. Energy Storage Materials. 2024;**68**:103343. DOI: 10.1016/j.ensm.2024.103343

[50] Wang Q, Ye M, Cai X, et al. Transferable data-driven capacity estimation for lithium-ion batteries with deep learning: A case study from laboratory to field applications. Applied Energy. 2023;**350**:121747. DOI: 10.1016/j.apenergy.2023.121747

[51] Geslin A, Xu L, Ganapathi D, et al. Dynamic cycling enhances battery lifetime. Nature Energy. 2024:1-9. DOI: 10.1038/s41560-024-01675-8

[52] Liang T, Glossner J, Wang L, et al. Pruning and quantization for deep neural network acceleration: A survey. Neurocomputing. 2021;**461**:370-403. DOI: 10.1016/j.neucom.2021.07.045

[53] Rokh B, Azarpeyvand A, Khanteymoori A. A comprehensive survey on model quantization for deep neural networks in image classification. ACM Transactions on Intelligent Systems and Technology. 2023;**14**:1-50. DOI: 10.1145/3623402

[54] Buber E, Banu D. Performance analysis and CPU vs GPU comparison for deep learning. In: 2018 6th International Conference on Control Engineering & Information Technology (CEIT). Piscataway, NJ, USA: IEEE; 2018. pp. 1-6. DOI: 10.1109/CEIT.2018.8751930

[55] Nair P, Vakharia V, Shah M, et al. AI-driven digital twin model for reliable lithium-ion battery discharge capacity predictions. International Journal of Intelligent Systems. 2024;**2024**(1):8185044. DOI: 10.1155/2024/8185044

[56] Zhao J, Wang Z, Wu Y, et al. Predictive pretrained transformer (PPT) for real-time battery health diagnostics. Applied Energy. 2025;**377**:124746. DOI: 10.1016/j.apenergy.2024.124746

[57] Xu R, Wang Y, Chen Z. A hybrid approach to predict battery health combined with attention-based transformer and online correction. Journal of Energy Storage. 2023;**65**:107365. DOI: 10.1016/j.est.2023.107365

[58] Hong J, Lee D, Jeong ER, et al. Towards the swift prediction of the remaining useful life of lithium-ion batteries with end-to-end deep learning. Applied Energy. 2020;**278**:115646. DOI: 10.1016/j.apenergy.2020.115646

[59] Li P, Zhang Z, Grosu R, et al. An end-to-end neural network framework for

state-of-health estimation and remaining useful life prediction of electric vehicle lithium batteries. Renewable and Sustainable Energy Reviews. 2022;**156**:111843. DOI: 10.1016/j.rser.2021.111843

[60] You G, Park S, Oh D. Real-time state-of-health estimation for electric vehicle batteries: A data-driven approach. Applied Energy. 2016;**176**:92-103. DOI: 10.1016/j.apenergy.2016.05.051

[61] Ma G, Xu S, Jiang B, et al. Real-time personalized health status prediction of lithium-ion batteries using deep transfer learning. Energy & Environmental Science. 2022;**15**(10):4083-4094. DOI: 10.1039/D2EE01676A

[62] Gu X, See KW, Li P, et al. A novel state-of-health estimation for the lithium-ion battery using a convolutional neural network and transformer model. Energy. 2023;**262**:125501. DOI: 10.1016/j.energy.2022.125501

[63] Jia C, Tian Y, Shi Y, et al. State of health prediction of lithium-ion batteries based on bidirectional gated recurrent unit and transformer. Energy. 2023;**285**:129401. DOI: 10.1016/j.energy.2023.129401

[64] Tang A, Xu Y, Hu Y, et al. Battery state of health estimation under dynamic operations with physics-driven deep learning. Applied Energy. 2024;**370**:123632. DOI: 10.1016/j.apenergy.2024.123632

[65] Hasani R, Lechner M, Amini A, et al. Liquid time-constant networks. In: Proceedings of the AAAI Conference on Artificial Intelligence. Vol. 35(9). Washington, DC, USA: AAAI; 2021. pp. 7657-7666. DOI: 10.1609/aaai.v35i9.16936

[66] Chahine M, Hasani R, Kao P, et al. Robust flight navigation out of distribution with liquid neural networks. Science. Robotics. 2023;**8**(77):eadc8892. DOI: 10.1126/scirobotics.adc8892

[67] Kairouz P, McMahan HB, Avent B, et al. Advances and open problems in federated learning. Foundations and Trends® in Machine Learning. 2021;**14**(1-2):1-210. DOI: 10.1561/2200000083

[68] Wong KL, Tse R, Tang SK, et al. Decentralized deep learning approach for lithium-ion batteries state of health forecasting using federated learning. IEEE Transactions on Transportation Electrification. 2024;**10**:8199-8212. DOI: 10.1109/TTE.2024.3354551

[69] Kröger T, Belnarsch A, Bilfinger P, et al. Collaborative training of deep neural networks for the lithium-ion battery aging prediction with federated learning. eTransportation. 2023;**18**:100294. DOI: 10.1016/j.etran.2023.100294

[70] He X, Zhao K, Chu X. AutoML: A survey of the state-of-the-art. Knowledge-Based Systems. 2021;**212**:106622. DOI: 10.1016/j.knosys.2020.106622

[71] Jin H, Cui N, Cai L, et al. State-of-health estimation for lithium-ion batteries with hierarchical feature construction and auto-configurable Gaussian process regression. Energy. 2023;**262**:125503. DOI: 10.1016/j.energy.2022.125503

[72] Karmaker SK, Hassan MM, Smith MJ, et al. Automl to date and beyond: Challenges and opportunities. ACM Computing Surveys (CSUR). 2021;**54**(8):1-36. DOI: 10.1145/3470918

Chapter 9

Application of Machine Learning for Vehicle Technologies on the Example of Diesel Engine Optimization

Hasan Koten and Kubilay Guner

Abstract

This study presents a machine learning-based framework to optimize internal combustion engine designs using data derived from multidimensional simulations. Traditional approaches to engine analysis and optimization often require significant time and computational resources, potentially spanning months or even years. By employing machine learning, this research provides a means to model complex physical processes within internal combustion engines, allowing for accurate predictions based on mathematical and statistical analyses of simulation data. The focus of this study is on the common rail injector parameters of a heavy-duty engine. Key input parameters for the simulations include in-cylinder flow dynamics, thermodynamic conditions at fuel injection, injection pressure, temperature, and intake valve closure timing. The outputs consist of performance and emissions benchmarks, which are then utilized to train machine learning models capable of predicting target metrics based on these inputs. A variety of machine learning algorithms were employed to build predictive models, demonstrating their ability to replicate the results of detailed simulations efficiently. This approach aims to accelerate the optimization process, reducing the need for exhaustive simulations while maintaining high accuracy. The findings highlight significant time and cost savings compared to traditional methods, showcasing the potential of machine learning to transform engine design and optimization processes.

Keywords: artificial intelligence, machine learning, deep learning, diesel engine, vehicle technology

1. Introduction

With new regulations made in line with consumer demands, companies aim to find new engine designs that simultaneously reduce fuel consumption and emissions. Research and development centers and companies use experimental prototyping and numerical modeling to develop these new designs. With the developing structures, numerical simulation can be done much more comprehensively than previous studies. Therefore, engine design optimization has gained much greater importance.

The investigation of the effects of nanostructured fuel droplets in diesel engines with a machine learning approach and rapid optimization technique is a complex and multi-disciplinary research area that requires expertise in mechanical engineering, materials science, and data science.

The main objective of this research is to explore the potential of using nanostructured fuel droplets to improve the efficiency and reduce emissions of diesel engines. Nanostructured fuel droplets are expected to have several advantages over traditional fuel droplets, including better atomization and combustion, higher energy density, and reduced emissions.

Machine learning techniques can be used to analyze large datasets of engine performance data, as well as simulations and experimental data, to identify patterns and relationships between different variables. This can help to optimize engine performance and identify the most promising nanostructured fuel droplets for further study.

Rapid optimization techniques, such as genetic algorithms and simulated annealing, can be used to quickly explore the design space of different nanostructured fuel droplets and identify optimal configurations. These techniques can help to reduce the time and cost of the experimental testing required to validate the performance of different fuel droplet designs.

Overall, this research has the potential to make significant contributions to the development of more efficient and environmentally-friendly diesel engines. However, it will require a collaborative effort between experts in different fields and the use of advanced computational tools and techniques.

The investigation of the effects of nanostructured fuel droplets in diesel engines with a machine learning approach and rapid optimization technique can be a complex and multi-faceted problem, requiring a combination of expertise in engineering, physics, chemistry, and computer science. However, it can provide a general overview of the steps involved in such an investigation.

Define the problem: The first step in any research project is to define the problem and establish clear research questions. In this case, the problem is to investigate the effects of using nanostructured fuel droplets in diesel engines, and the research questions could include: What are the effects on engine performance, emissions, and fuel consumption? What are the optimal conditions for using nanostructured fuel droplets in diesel engines? What is the potential for scaling up this technology?

Data collection: In order to investigate the effects of nanostructured fuel droplets, data must be collected from experiments or simulations. This data could include information on engine performance, emissions, fuel consumption, fuel properties, and the properties of the nanostructured fuel droplets. Machine learning algorithms require a large amount of data, so it is important to ensure that the data is reliable and representative.

Data preprocessing: Before the data can be used for machine learning, it must be preprocessed to remove outliers, normalize the data, and ensure that it is in the correct format for the machine learning algorithms. This step is crucial for ensuring that the machine learning algorithms can accurately analyze the data.

Machine learning: Once the data is preprocessed, it can be used to train machine learning algorithms. These algorithms can be used to analyze the data and identify patterns and relationships between the different variables. The machine learning algorithms can also be used to develop predictive models that can be used to optimize the use of nanostructured fuel droplets in diesel engines.

Rapid optimization: Once the machine learning algorithms have developed predictive models, these models can be used in rapid optimization techniques. Rapid optimization techniques involve using computer simulations to test different variables and identify the optimal conditions for using nanostructured fuel droplets in diesel engines. This process can be used to identify the optimal droplet size, injection timing, and other variables that can affect engine performance, emissions, and fuel consumption.

Validation: Once the optimal conditions have been identified, it is important to validate these conditions using experiments or simulations. This step is crucial for ensuring that the predictive models developed by the machine learning algorithms are accurate and reliable.

In summary, investigating the effects of nanostructured fuel droplets in diesel engines with a machine learning approach and rapid optimization technique requires a combination of data collection, preprocessing, machine learning, rapid optimization, and validation. This approach can provide valuable insights into the optimal use of nanostructured fuel droplets in diesel engines and can help to improve engine performance, reduce emissions, and increase fuel efficiency.

Studies cover a range of topics, including an overview of different machine learning techniques that can be used for engine calibration, case studies on the use of these techniques for engine parameter modeling, optimization, and prediction of emissions, and novel approaches to engine performance optimization. Overall, these articles highlight the potential benefits of using machine learning for engine calibration and optimization, such as improved performance, reduced emissions, and reduced development time and costs. They also emphasize the need for further research in this area to fully realize these benefits and advance the field of engine optimization [1–6].

The articles referenced focus on the use of advanced machine learning techniques for modeling diesel engine performance under limited or scarce data sets. The techniques used include artificial neural networks, genetic algorithms, deep learning, support vector regression, and random forest. These approaches demonstrate the potential to accurately predict engine performance under limited or scarce data conditions, which can be useful for engine design and optimization. Overall, the articles highlight the importance of utilizing advanced machine learning techniques to improve the accuracy of diesel engine performance modeling under limited or scarce data conditions [7–12].

In studies note that internal combustion engines are complex systems with many variables that affect their performance, including factors such as fuel injection timing, ignition timing, and air-fuel ratio. Traditional methods for optimizing engine performance typically involve extensive experimentation and testing, which can be time-consuming and expensive. They note that machine learning algorithms can identify patterns in the data that might not be apparent to human operators, allowing for more accurate optimization of engine performance [13].

The articles demonstrate the use of machine learning techniques such as artificial neural networks, Gaussian process modeling, support vector machines, deep reinforcement learning, and hybrid approaches to optimize engine performance. Overall, the articles suggest that machine learning approaches can be effective in optimizing engine performance, and further research in this area could lead to even more significant improvements in engine design and operation [14–16].

The articles provides a detailed overview of the various machine learning techniques that can be used for engine optimization, including linear regression, decision trees, and artificial neural networks. The authors conclude that machine learning techniques have great potential for optimizing engine performance and reducing the time and costs associated with traditional optimization methods. They note that further research is needed to fully realize the potential of machine learning for engine optimization, but that the approach shows promise for improving the efficiency and performance of internal combustion engines [17, 18].

In some studies demonstrate the use of kernel-based extreme learning machine, cuckoo search, particle swarm optimization, support vector regression, and gray wolf optimizer in modeling and optimizing biodiesel engine performance. The articles highlight the potential benefits of using these techniques, including improved fuel efficiency, reduced emissions, and reduced development time and costs. Overall, the articles suggest that these approaches can be effective in modeling and optimizing biodiesel engine performance, and further research in this area could lead to even more significant improvements in biodiesel engine design and operation [19, 20]. The main goal of these studies is to improve the fuel efficiency and reduce emissions of the biodiesel engine, which can ultimately lead to a more sustainable and environmentally-friendly transportation system. Additionally, all of these studies focus on the optimization of biodiesel engine performance, which indicates a growing interest in the use of alternative fuels in the transportation industry [21–23].

Some studies the optimization of biodiesel engine performance and emissions using a hybrid genetic algorithm and particle swarm optimization algorithm. The main goal is to improve the fuel efficiency and reduce emissions of the biodiesel engine. The studies use different variations of the hybrid algorithm, such as enhanced hybrid particle swarm optimization (PSO) – genetic algorithm (GA) and hybrid GA-PSO, and apply them to different biodiesel engines. Additionally, some studies also explore multi-objective optimization of engine performance and emissions. Overall, these studies suggest that the hybrid algorithm approach can effectively optimize the performance and emissions of biodiesel engines, and the enhanced hybrid PSO-GA algorithm may offer better results compared to basic GA or hybrid GA-PSO algorithms [24–29].

Some studies have also been compared. Presents an experimental study that measures the start of combustion timing in a methane-fueled HCCI engine and compares the results with predictions made by traditional methods, such as the heat release rate analysis, and machine learning methods, such as artificial neural networks and support vector regression. The study aims to determine which method is more accurate and reliable for predicting the start of combustion timing in methane-fueled HCCI engines. The results show that the machine learning methods, particularly support vector regression, outperform the traditional methods in terms of accuracy and reliability. This suggests that machine learning methods can be a valuable tool for predicting the start of combustion in HCCI engines, which can ultimately lead to better engine performance and emissions control [30].

Studies on ship diesel engines provide an introduction and overview of the use of hybrid models that combine computational fluid dynamics (CFD) simulations and machine learning algorithms to predict the performance of ship diesel engines. The authors discuss the importance of accurate performance predictions for optimizing engine design and operation, as well as the challenges involved in developing models that can capture the complex physics of engine operation [31–33]. The authors focus on the development and validation of hybrid models that combine CFD simulations and machine learning algorithms to predict various engine performance parameters. Also describe the different types of inputs and parameters that can be included in the models, such as engine speed, load, and fuel type, and discuss the challenges involved in selecting appropriate inputs and parameters. They in addition describe the different machine learning algorithms that can be used for modeling engine performance, such as neural networks and support vector machines, and compare their effectiveness. Finally, the authors validate their models using experimental data and demonstrate their accuracy in predicting engine performance [34–37]. They also describe the different techniques that can be used to optimize the models, such as sensitivity analysis and feature selection, and demonstrate their effectiveness in improving model accuracy [38, 39]. They focus specifically on the prediction of engine combustion and emissions using hybrid models. They also demonstrate the effectiveness of hybrid models in predicting engine combustion and emissions and provide insights into the factors that affect these parameters [40, 41].

Studies provide an introduction and overview of Atkinson cycle gasoline engines and their advantages in terms of fuel efficiency and emissions reduction. They describe the basic principles of the Atkinson cycle, which involves a longer expansion stroke than a compression stroke, and compare it with the conventional Otto cycle. The authors also discuss the different approaches that can be used to optimize the performance of Atkinson cycle engines, such as response surface methodology and genetic algorithms, and provide insights into the factors that affect engine performance. He et al. focus specifically on the experimental analysis of Atkinson cycle gasoline engines. The authors describe the different types of experiments that can be performed to evaluate engine performance, such as brake power, fuel consumption, and emissions measurements, and provide a detailed analysis of the experimental results. They also investigate the effects of different engine parameters, such as the compression ratio and valve timing, on engine performance and provide insights into the underlying physics of the Atkinson cycle [42, 43]. Articles focus on the computational optimization of Atkinson cycle gasoline engines using various optimization algorithms, including machine learning and genetic algorithms. The authors describe the different approaches that can be used to optimize engine performance, such as multi-objective optimization and response surface methodology, and compare the effectiveness of these methods. They also investigate the effects of different engine parameters on engine performance and use optimization algorithms to identify the optimal values of these parameters. Finally, the authors validate their optimization results using experimental data and demonstrate the effectiveness of their optimization methods in improving engine performance [44–46].

Also, these articles focus on the use of machine learning techniques for high-performance engine calibration. They describe different machine learning algorithms, such as artificial neural networks, support vector machines, and random forests, and how they can be applied to the calibration process. The authors discuss the advantages of using machine learning techniques, such as increased accuracy and reduced calibration time, and demonstrate the effectiveness of their methods through experimental validation. They also investigate the impact of different factors on engine performance, such as fuel type and engine load, and how machine learning techniques can be used to optimize engine calibration for different operating conditions [47–51].

The studies provide an introduction and overview of the use of machine learning and genetic algorithms in optimization processes. They describe the basic principles of machine learning and genetic algorithms, and their applications in different fields. They focus specifically on the optimization of specific processes, such as predicting the mechanical properties of carbon fiber composites and optimizing cutting force in end milling processes. They also discuss the challenges involved in optimizing these processes and the advantages of using machine learning and genetic algorithm techniques in these applications. Finally, the studies demonstrate the effectiveness of these techniques in optimizing the specific processes [52, 53].

In other studies focus on the use of the ML-GA approach for optimization processes in different fields. The authors describe the different machine learning and genetic algorithm techniques used in their studies and how they were combined to form the ML-GA approach. They also discuss the challenges involved in optimizing complex systems and the advantages of using the ML-GA approach, such as reduced optimization time and increased accuracy. The studies demonstrate the effectiveness of the ML-GA approach in optimizing different systems, such as multi-pass turning processes, dual fuel diesel engines, and placement of solar water heaters [54–56].

The integration of machine learning (ML) techniques in analyzing fuel droplets in diesel engines has shown significant promise in enhancing the understanding of spray processes, droplet behavior, and fuel consumption predictions. By leveraging advanced ML models, researchers can quantify droplet size distributions and improve the accuracy of simulations related to droplet heating and evaporation, which are critical for optimizing combustion efficiency.

Machine Learning in Spray Characterization and ML object detection methods, such as Mask R-CNN and SparseInst, have been employed to analyze droplet size and primary atomization processes in diesel sprays [57]. These models utilize synthetic training data and physics-informed domain randomization, achieving robust detection across various measurement conditions.

Droplet Heating and Evaporation Models: The multidimensional quasi-discrete (MDQD) model effectively simulates the heating and evaporation of diesel fuel droplets, accounting for the complex composition of diesel [58]. This model significantly reduces computational time while maintaining accuracy, demonstrating the importance of precise droplet behavior modeling in engine simulations.

Predictive Modeling for Fuel Consumption: AI techniques have been utilized to identify optimal ML models for predicting fuel consumption, with Gaussian process regression (GPR) outperforming other models in accuracy and speed [59]. Sensitivity analysis within these models highlights key parameters influencing fuel consumption, enhancing predictive reliability.

While ML applications in diesel engine research are promising, challenges remain in ensuring model generalizability across diverse operating conditions and fuel compositions. Further research is needed to refine these models for broader applicability in real-world scenarios.

The integration of nano droplets in diesel engines, particularly through the use of nanoparticles, has shown significant potential in enhancing engine performance and reducing emissions. Machine learning techniques are increasingly being employed to optimize these processes, allowing for more precise control and prediction of engine behavior under various conditions. The following sections outline the key aspects of this research area.

Performance Enhancement: Nanoparticle Blends: The use of hybrid nanoparticles, such as MgO and Al2O3, in water-in-diesel emulsions has been shown to improve brake thermal efficiency and reduce brake-specific fuel consumption [60]. Optimization Techniques, Response Surface Methodology, and Genetic Algorithms have been utilized to identify optimal operating parameters, achieving a brake thermal efficiency with minimized emissions [61].

DOI: http://dx.doi.org/10.5772/intechopen.1010179

Emission Reduction: Catalytic Effects: The addition of carbon nanotubes to biodiesel has led to significant reductions in CO, NO, and NO_2 emissions, demonstrating the effectiveness of nanoparticles in mitigating harmful exhaust outputs. Comparative Analysis: Studies indicate that nanofuels generally produce lower CO and HC emissions while increasing NOx emissions, highlighting the need for balanced formulations [62].

Machine Learning Applications: Predictive Modeling: Machine learning models, such as deep neural networks, have been employed to predict engine performance and emissions based on various input parameters, enhancing the accuracy of experimental outcomes [63].

While the advancements in nanodroplet technology and machine learning present promising avenues for improving diesel engine efficiency and sustainability, challenges such as production costs and long-term fuel stability remain critical barriers to widespread adoption [64].

2. Materials and methods

This study employs a machine learning-based framework to optimize the design parameters of internal combustion engines using data derived from multidimensional simulations. The primary focus is on the optimization of common rail injector parameters in a heavy-duty diesel engine, aiming to enhance performance and reduce emissions.

Data for this study were obtained through high-fidelity simulations, incorporating key engine parameters such as in-cylinder flow dynamics, thermodynamic conditions at fuel injection, injection pressure, temperature, and intake valve closure timing. These input variables were selected based on their critical influence on engine performance and emissions. The output parameters include performance benchmarks such as brake thermal efficiency, specific fuel consumption, and emissions levels (CO, NOx, and particulate matter). Prior to training the machine learning models, data preprocessing steps were implemented, including normalization, outlier detection, and feature engineering. These preprocessing steps ensured that the data were in a suitable format for accurate and efficient model training. A range of machine learning algorithms were evaluated for their predictive accuracy, including artificial neural networks (ANN), support vector regression (SVR), random forests (RF), and deep learning models. The models were trained using historical simulation data, with hyperparameter tuning performed to optimize predictive performance. Cross-validation techniques were applied to prevent overfitting and ensure model robustness.

2.1 Optimization techniques

To expedite the optimization process, rapid optimization techniques such as genetic algorithms (GA) and particle swarm optimization (PSO) were employed. These techniques efficiently explored the parameter space to identify optimal engine configurations that maximize efficiency while minimizing emissions. The integration of machine learning with optimization algorithms significantly reduced the computational costs and time required for traditional trial-and-error approaches. The trained models were validated against experimental and additional simulation data to assess their generalizability and accuracy. Performance metrics such as mean squared error

(MSE), R-squared (R^2), and root mean square error (RMSE) were used to evaluate model predictions against ground truth values.

3. Results and discussion

The findings of this study demonstrate the potential of machine learning in optimizing internal combustion engine performance with reduced computational costs. The predictive models developed in this research successfully captured complex nonlinear relationships between input parameters and engine performance indicators. The integration of machine learning with optimization algorithms provided a systematic approach to engine design, enabling the rapid identification of optimal operating conditions. One of the key insights from this study is the ability of machine learning models to replicate the results of high-fidelity simulations with high accuracy. Traditional engine optimization methods often require extensive computational resources and time, making them impractical for real-time applications. In contrast, machine learning models provide a feasible alternative by reducing the number of simulations needed to achieve optimized performance. The application of genetic algorithms and particle swarm optimization further enhanced the efficiency of the optimization process. These techniques enabled rapid exploration of the design space, identifying configurations that optimize fuel efficiency and emissions reduction. The results highlight the advantages of using AI-driven approaches over conventional experimental methods, particularly in terms of cost-effectiveness and time efficiency. However, challenges remain in ensuring model generalizability across different engine types and operating conditions. While the trained models performed well within the dataset used for validation, their performance in real-world applications may require further fine-tuning and additional experimental validation. Moreover, the interpretability of complex machine learning models remains a challenge, necessitating the integration of explainable AI techniques to enhance transparency and trust in model predictions. Future research should focus on expanding the dataset to include real-world driving conditions and additional variables influencing engine performance. Additionally, hybrid modeling approaches that combine physics-based simulations with machine learning could further improve predictive accuracy and robustness. The incorporation of reinforcement learning and adaptive algorithms could also enable real-time optimization of engine parameters, paving the way for intelligent, self-optimizing engine control systems. Overall, this study underscores the transformative potential of machine learning in vehicle technology optimization. By leveraging AI-driven methods, the automotive industry can achieve more efficient, cost-effective, and environmentally sustainable engine designs, contributing to advancements in both performance and emissions reduction.

4. Conclusion

It is seen that significant contributions by leveraging machine learning approaches to accelerate the analysis and optimization processes of internal combustion engines. Specifically, simulations were conducted for the common rail injector parameters of a heavy-duty engine, with input parameters including various in-cylinder flow and thermodynamic conditions at fuel injection, pressure, temperature, and intake valve closure. The outputs from these simulations included benchmarks related to engine performance and emissions. The use of machine learning models has significantly reduced the number of simulations required for the optimization process, compared to traditional methods. By applying various machine learning algorithms, the study achieved cost savings and accelerated optimization without compromising accuracy. In conclusion, this research has demonstrated the effective use of machine learning techniques in the optimization of internal combustion engines, providing a foundation for future studies in this field. This approach enables the development of faster and more cost-effective solutions in engineering and design processes. These findings can be considered an important step toward enhancing the performance and efficiency of internal combustion engines in the automotive industry.

Acknowledgements

The authors acknowledge the use of openAI tool for language polishing of the manuscript.

Author details

Hasan Koten[1]* and Kubilay Guner[2]

1 Mechanical Engineering Department, Istanbul Medeniyet University, Istanbul, Turkiye

2 Institute of Graduate Studies, Istanbul Medeniyet University, Istanbul, Turkiye

*Address all correspondence to: hasan.koten@medeniyet.edu.tr

IntechOpen

© 2025 The Author(s). Licensee IntechOpen. This chapter is distributed under the terms of the Creative Commons Attribution License (http://creativecommons.org/licenses/by/4.0), which permits unrestricted use, distribution, and reproduction in any medium, provided the original work is properly cited.

References

[1] Van Horn AK. Machine Learning Techniques for High Performance Engine Calibration. 2008. Available from: https://hdl.handle.net/1808/14610

[2] Amalyan E, Latifi S. Machine learning-based grading of engine health for high-performance vehicles. Electronics. 2025;**14**(3):475. DOI: 10.3390/electronics14030475

[3] Ishola NB, Epelle EI, Betiku E. Machine learning approaches to modeling and optimization of biodiesel production systems: State of art and future outlook. Renewable and Sustainable Energy Reviews. 2024;**152**:111620. DOI: 10.1016/j.ecmx.2024.100669

[4] Wong PK, Gao XH, Wong KI, Vong CM. Efficient point-by-point engine calibration using machine learning and sequential design of experiment strategies. Journal of the Franklin Institute. 2018;**355**(4):1517-1538. DOI: 10.1016/j.jfranklin.2017.02.006

[5] García A, Monsalve-Serrano J, Marco-Gimeno J, Iñiguez E. Data-driven modeling for predicting the steady-state and transient performance of a dual-fuel medium-duty engine employing artificial neural networks. Fuel. 2025;**394**:135150. DOI: 10.1016/j.fuel.2025.135150

[6] Ricci F, Avana M, Mariani F. A deep learning method for the prediction of pollutant emissions from internal combustion engines. Applied Sciences. 2024;**14**(21):9707. DOI: 10.3390/app14219707S

[7] Wong KI, Wong PK, Cheung CS, Vong CM. Modelling of diesel engine performance using advanced machine learning methods under scarce and exponential data set. Applied Soft Computing. 2013;**13**(11):4428-4441. DOI: 10.1016/j.asoc.2013.06.006

[8] Odufuwa OY, Tartibu LK, Kusakana K. Artificial neural network modelling for predicting efficiency and emissions in mini-diesel engines: Key performance indicators and environmental impact analysis. Fuel. 2025;**387**:134294. DOI: 10.1016/j.fuel.2025.134294

[9] He Y, Rutland CJ. Modeling of a Turbocharged DI Diesel Engine Using Artificial Neural Networks. SAE Technical Paper, 2002-01-2772. 2002. pp. 1-11. DOI: 10.4271/2002-01-2772

[10] Khanyi N, Inambao FL, Stopforth R. A comprehensive review of the GT-POWER for modelling diesel engines. Energies. 2025;**18**(8):1880. DOI: 10.3390/en18081880

[11] Wong KI, Wong PK, Cheung CS. Modelling and prediction of diesel engine performance using relevance vector machine. International Journal of Green Energy. 2015;**12**(3):265-271. DOI: 10.1080/15435075.2014.891513

[12] Kumar KS, Razak A, Ramis MK, Irshad SM, Islam S, Wodajo AW. Statistical and machine learning analysis of diesel engines fueled with Moringa oleifera biodiesel doped with 1-hexanol and Zr_2O_3 nanoparticles. Scientific Reports. 2025;**15**:7269. DOI: 10.1038/s41598-025-87818-7

[13] Dutta P, Sharma S, Rathnam PA. Engine performance optimization using machine learning techniques. In: Proceedings of the 2015 SAI Intelligent Systems Conference (IntelliSys). pp. 620-625. DOI: 10.1109/IntelliSys.2015.7361134

[14] Kesgin U. Genetic algorithm and artificial neural network for engine optimisation of efficiency and NO_x emission. Fuel. 2004;**83**(7-8):885-895. DOI: 10.1016/j.fuel.2003.10.025

[15] Shi X, Jiang D, Qian W, Liang Y. Application of the Gaussian process regression method based on a combined kernel function in engine performance prediction. ACS Omega. 2022;**7**(45):41732-41743. DOI: 10.1021/acsomega.2c05952

[16] He P, Chen J-H, Zhang C-Z, Yu Z-Y, Wang M-Y, Chen J-Y, et al. Optimization of modeling and temperature control of air-cooled PEMFC based on TLBO–DE. Applied Thermal Engineering. 2025;**276**:126841. DOI: 10.1016/j.applthermaleng.2025.126841

[17] Wang Y, Wang G, Yao G, Shen Q, Yu X, He S. Combining GA-SVM and NSGA-III multi-objective optimization to reduce the emission and fuel consumption of high-pressure common-rail diesel engine. Energy. 2023;**278**:127965. DOI: 10.1016/j.energy.2023.127965

[18] Sehgal A, La HM, Louis SJ, Nguyen H. Deep reinforcement learning using genetic algorithm for parameter optimization. arXiv preprint, 1905.04100. 2019. DOI: 10.48550/arXiv.1905.04100

[19] Wong PK, Wong KI, Vong CM, Cheung CS. Modeling and optimization of biodiesel engine performance using kernel-based extreme learning machine and cuckoo search. Renewable Energy. 2015;**74**:640-647. DOI: 10.1016/j.renene.2014.08.075

[20] Şahin S, Torun A. Comparison of engine performance and emission values of biodiesel obtained from waste pumpkin seeds with machine learning. Agriculture. 2024;**14**(2):227. DOI: 10.3390/agriculture14020227

[21] Liang Y, Niu D, Ye M, Hong W-C. Short-term load forecasting based on wavelet transform and least squares support vector machine optimized by improved cuckoo search. Energies. 2016;**9**:827. DOI: 10.3390/en9100827

[22] Shirneshan A, Hosseinzadeh Samani B, Ghobadian B. Optimization of biodiesel percentage in fuel mixture and engine operating conditions for diesel engine performance and emission characteristics by artificial bees colony algorithm. Fuel. 2016;**184**:518-526. DOI: 10.1016/j.fuel.2016.06.117

[23] Agrawal P, Gnanaprakash R, Dhawane SH. Prediction of biodiesel yield employing machine learning: Interpretability analysis via Shapley additive explanations. Fuel. 2024;**359**:130516. DOI: 10.1016/j.fuel.2023.130516

[24] Zhang Q, Ogren RM, Kong S-C. A comparative study of biodiesel engine performance optimization using enhanced hybrid PSO–GA and basic GA. Applied Energy. 2016;**165**:676-684. DOI: 10.1016/j.apenergy.2015.12.044

[25] Ahmad A, Yadav AK, Singh A, Singh DK, Ağbulut Ü. A hybrid RSM–GA–PSO approach on optimization of process intensification of linseed biodiesel synthesis using an ultrasonic reactor: Enhancing biodiesel properties and engine characteristics with ternary fuel blends. Energy. 2024;**288**:129077. DOI: 10.1016/j.energy.2023.129077

[26] Şen M. Optimization of performance and emission of a diesel engine fueled with isopropyl alcohol blends: A comparative ANN-GA and RSM–HCO application. Engineering Science and Technology, an International Journal. 2024;**55**:101733. DOI: 10.1016/j.jestch.2024.101733

[27] Leong CC, Blakey S, Wilson CW. Genetic algorithm optimised chemical

reactors network: A novel technique for alternative fuels emission prediction. Swarm and Evolutionary Computation. 2016;**27**:180-187. DOI: 10.1016/j.swevo.2015.12.001

[28] Bendu H, Deepak BBVL, Murugan S. Multi-objective optimization of ethanol fuelled HCCI engine performance using hybrid GRNN–PSO. Applied Energy. 2017;**187**:601-611. DOI: 10.1016/j.apenergy.2016.11.072

[29] Veza I, Karaoglan AD, İleri E, Afzal A, Hoang AT, Tamaldin N, et al. Multi-objective optimization of diesel engine performance and emission using grasshopper optimization algorithm. Fuel. 2022;**323**:124303. DOI: 10.1016/j.fuel.2022.124303

[30] Namar MM, Jahanian O, Köten H. The start of combustion prediction for methane-fueled HCCI engines: Traditional vs. machine learning methods. Mathematical Problems in Engineering. 2022;**2022**:Article 4589160, 1-11. DOI: 10.1155/2022/4589160

[31] Coraddu A, Kalikatzarakis M, Oneto L, Meijn GJ, Godjevac M, Geertsma RD. Ship diesel engine performance modelling with combined physical and machine learning approach. In: Proceedings of the International Ship Control Systems Symposium. Vol. 1. 2018. DOI: 10.24868/issn.2631-8741.2018.011

[32] Vorkapić A, Radonja R, Babić K, Martinčić-Ipšić S. Machine learning methods in monitoring operating behaviour of marine two-stroke diesel engine. Transport. 2020;**35**(5):474-485. DOI: 10.3846/transport.2020.14038

[33] Lee J, Eom J, Park J, Jo J, Kim S. The development of a machine learning-based carbon emission prediction method for a multi-fuel-propelled smart ship by using onboard measurement data. Sustainability. 2024;**16**(6):2381. DOI: 10.3390/su16062381

[34] Shahpouri S, Norouzi A, Hayduk C, Rezaei R, Shahbakhti M, Koch CR. Hybrid machine learning approaches and a systematic model selection process for predicting soot emissions in compression ignition engines. Energies. 2021;**14**(23):7865. DOI: 10.3390/en14237865

[35] Badra JA, Khaled F, Tang M, Pei Y, Kodavasal J, Pal P, et al. Engine combustion system optimization using computational fluid dynamics and machine learning: A methodological approach. Journal of Energy Resources Technology, Transactions of the ASME. 2021;**143**(2):Article 022306. DOI: 10.1115/1.4047978

[36] Mohammad A, Rezaei R, Hayduk C, Delebinski TO, Shahpouri S, Shahbakhti M. Hybrid Physical and Machine Learning-Oriented Modeling Approach to Predict Emissions in a Diesel Compression Ignition Engine. SAE Technical Paper, 2021-01-0496. 2021. pp. 1-13. DOI: 10.4271/2021-01-0496

[37] Karystinos V, Papalambrou G. Deep long short-term memory neural networks as virtual sensors for marine diesel engine NO_x prediction at transient conditions. International Journal of Engine Research. 2024;**25**(5):1024-1036. DOI: 10.1177/14680874231217342

[38] Larsen U, Pierobon L, Baldi F, Haglind F, Ivarsson A. Development of a model for the prediction of the fuel consumption and nitrogen oxides emission trade-off for large ships. Energy. 2015;**80**:545-555. DOI: 10.1016/j.energy.2014.12.009

[39] Yüksel O, Bayraktar M, Sokukcu M. Comparative study of machine learning techniques to predict fuel consumption of a marine diesel engine. Ocean

Engineering. 2023;**286**:115505. DOI: 10.1016/j.oceaneng.2023.115505

[40] Chen X, Liu L, Du J, Liu D, Huang L, Li X. Intelligent optimization based on a virtual marine diesel engine using GA-ICSO hybrid algorithm. Machines. 2022;**10**(4):227. DOI: 10.3390/machines10040227

[41] Wakes S, Bauer BO, Mayo MA. A preliminary assessment of machine learning algorithms for predicting CFD-simulated wind flow patterns over idealised foredunes. Journal of the Royal Society of New Zealand. 2021;**51**(2):290-306. DOI: 10.1080/03036758.2020.1868541

[42] Qi L, Guo T, Fu J, Dai H, Liu J. Experimental study on the effects of injection parameters and exhaust gas recirculation on combustion, emission and performance of an Atkinson cycle gasoline direct-injection engine. Energy. 2022;**238**:121784. DOI: 10.1016/j.energy.2021.121784

[43] Tong J, Liang Y. Experiment analysis and computational optimization of the Atkinson cycle gasoline engine through NSGA II algorithm using machine learning. Energy Conversion and Management. 2021;**238**:113871. DOI: 10.1016/j.enconman.2021.113871

[44] Niu Q, Sun B, Zhang D, Luo Q. Research on performance optimization and fuel-saving mechanism of an Atkinson cycle gasoline engine at low speed and part load. Fuel. 2020;**265**:117010. DOI: 10.1016/j.fuel.2020.117010

[45] Feng R, Li Y, Yang J, Fu J, Zhang D, Zheng G. Investigations of Atkinson Cycle Converted from Conventional Otto Cycle Gasoline Engine. SAE Technical Paper, 2016-01-0680. 2016. pp. 1-8. DOI: 10.4271/2016-01-0680

[46] Zhao J, Xu M, Li M, Wang B, Liu S. Design and optimization of an Atkinson cycle engine with the artificial neural network method. Applied Energy. 2012;**92**:492-502. DOI: 10.1016/j.apenergy.2011.11.060

[47] Cho Y, Huber T, Lauff U, Reddy R. Optimisation of gasoline engines automation and machine learning techniques in calibration. ATZelektronik Worldwide. 2017;**12**:48-53. DOI: 10.1007/s38314-017-0031-4

[48] Azmin FM, Stobart R, Rutledge J, Winward E. Using a Statistical Machine Learning Tool for Diesel Engine Air Path Calibration. SAE Technical Paper, 2014-01-2391. 2014. pp. 1-8. DOI: 10.4271/2014-01-2391

[49] Liu J, Huang Q, Ulishney C, Dumitrescu CE. Machine learning assisted prediction of exhaust gas temperature of a heavy-duty natural gas spark ignition engine. Applied Energy. 2021;**300**:117413. DOI: 10.1016/j.apenergy.2021.117413

[50] Garg P, Silvas E, Willems FPT. Potential of machine learning methods for robust performance and efficient engine control development. IFAC-PapersOnLine. 2021;**54**(10):189-195. DOI: 10.1016/j.ifacol.2021.10.162

[51] Godwin DJ, Varuvel EG, Martin ML. Prediction of combustion, performance, and emission parameters of ethanol-powered spark ignition engine using ensemble least squares boosting machine learning algorithms. Journal of Cleaner Production. 2023;**421**:138401. DOI: 10.1016/j.jclepro.2023.138401

[52] Sonawane SB, Sekhar R, Warke A, Thipse SS, Rairikar SD, Varma C. Experimental investigation and prediction of combustion parameters using machine learning in ethanol–gasoline blended engines. Journal of Engineering and Technological Sciences. 2025;**57**(1):27-47. DOI: 10.5614/j.eng.technol.sci.2025.57.1.3

[53] Gupta AK, Guntuku SC, Desu RK, Balu A. Optimisation of turning parameters by integrating genetic algorithm with support vector regression and artificial neural networks. The International Journal of Advanced Manufacturing Technology. 2015;**77**(1-4):331-339. DOI: 10.1007/s00170-014-6282-9

[54] Zhang F, Mei Y, Nguyen S, Zhang M. Survey on genetic programming and machine learning techniques for heuristic design in job shop scheduling. IEEE Transactions on Evolutionary Computation. 2024;**28**(1):147-167. DOI: 10.1109/TEVC.2023.3255246

[55] Wen Z, Wang Q, Zhao J, Li X. Multi-objective optimization of a diesel engine–ORC combined system integrating artificial neural network with genetic algorithm. Fuel. 2024;**371**:131981. DOI: 10.1016/j.fuel.2024.131981

[56] Atia DM, Fahmy FH, Ahmed NM, Dorrah HT. Optimal sizing of a solar water heating system based on a genetic algorithm for an aquaculture system. Mathematical and Computer Modelling. 2012;**55**(3-4):1436-1449. DOI: 10.1016/j.mcm.2011.10.022

[57] Jose B, Hampp F. Machine learning based spray process quantification. International Journal of Multiphase Flow. 2024;**172**:104702. DOI: 10.1016/j.ijmultiphaseflow.2023.104702

[58] Sazhin SS, Al Qubeissi M, Nasiri R, Gun'ko VM, Elwardany AE, Lemoine F, et al. A multi-dimensional quasi-discrete model for the analysis of diesel fuel droplet heating and evaporation. Fuel. 2014;**129**:238-266. DOI: 10.1016/j.fuel.2014.03.028

[59] Shateri A, Yang Z, Xie J. Utilizing artificial intelligence to identify an optimal machine learning model for predicting fuel consumption in diesel engines. Energy and AI. 2024;**16**:100360. DOI: 10.1016/j.egyai.2024.100360

[60] Patel KR, Dhiman VD. Effect of hybrid nanoparticles MgO and Al2O3 added water-in-diesel emulsion fueled diesel engine using hybrid deep neural network-based spotted hyena optimization. Heat Transfer. 2022;**51**:5748-5788. DOI: 10.1002/htj.22568

[61] Dewangan* A, Mallick A, Yadav AK, Ahmad A, Alqahtani D, Islam S. Combined effect of operating parameters and nanoparticles on the performance of a diesel engine: Response surface methodology-coupled genetic algorithm approach. ACS Omega. 2023;**8**:24586-24600. DOI: 10.1021/acsomega.3c02782

[62] Akhil P, Parashar S, Kolape P, Musthafa FM. Experimental investigation on performance, combustion and emission characteristics of nanoparticle blends with diesel as fuel in a CI engine. INCAS Bulletin. 2022;**14**:31-45. DOI: 10.13111/2066-8201.2022.14.3.4

[63] Beirami AAM, Maghsoudlou E, Nasrabadi M, Sergeevna KN, Abdullaev S, Ibrahim W. An assessment of greenhouse gases emission from diesel engine by adding carbon nanotube to biodiesel fuel using machine learning technique. Low-Carbon Technologies. 2024;**19**:1358-1367. DOI: 10.1093/ijlct/ctae076

[64] ul Haq M, Jafry T, Ali M, Ijaz Malik MA. Influence of nano additives on diesel-biodiesel fuel blends in diesel engine: A spray, performance, and emissions study. Energy Conversion and Management. 2024;**X**:100574. DOI: 10.1016/j.ecmx.2024.100574

Chapter 10

New Horizons in the Electromobility Industry: Future Job Roles Re-/Upskilling

Jesus M. Orona-Hinojos, Alejandro Yamaguchi Torres-Valdez, Lizbeth A. Huerta-Larrumbe, Eduardo Macias-Avila and Alfredo Rios-Nogues

Abstract

The electromobility industry is experiencing rapid transformation, driven by key technological shifts that drive the development of electric vehicles (EVs). This shift not only revolutionizes mobility but also creates new opportunities and challenges for the workforce. As the industry embraces electric mobility and autonomous driving technologies, new roles are emerging while existing jobs are evolving. The last few years have witnessed a fast growth in the field of self-driving vehicles, mainly backed up by advances in the area of machine learning and Generative AI Algorithms. As vehicles become increasingly complex, it is virtually impossible for car manufacturers to model how each component of the system should behave at any given time. Technology is now at a point where machines can be configured to learn from observation, just like humans.

Keywords: electric vehicles, set skills, machine learning (ML), data science, new job roles

1. Introduction

This writing represents an important challenge since it intends to relate, on the one hand, a social demand, such as the education of the workforce of the automotive industry, and on the other hand, the demand for sustainable mobility of EVs that converge in a framework of complex economics and social regulation in a post-pandemic and pre-climate era world.

During the transition of electromobility, the main drivers of workforce change include [1–4]:

- EV's new job roles: From battery production specialists and electric powertrain engineers to software developers and sustainability experts, the demand for specialized skills is rising. The transition to EVs requires a reimagined approach to vehicle design, production, and maintenance, creating diverse career paths in manufacturing, research, and tech development.

- EV's reskilling and upskilling: Traditional automotive workers must acquire new skills in electric vehicle technology, advanced manufacturing, data analytics, and AI. Companies and educational institutions are developing training programs and partnerships to equip workers with the expertise needed for EV production and related fields.

- EV's technological innovation assimilation: The rise of machine learning and autonomous driving is creating new challenges for workforce development. How to express Jean W. Piaget (1960): through the assimilation of new techniques or knowledge, we fit new experiences into our existing schemas. However, when new experiences do not fit, we adjust our schemas or create new ones through accommodation.

Therefore, in a chapter, we can only give a few outlines and outlines of what we face and give reasoned and reflective answers that can be applied under certain assumptions, seeking to answer questions similar to these in various forums:

- How are technological advancements in EVs impacting job roles in the automotive industry?

- What are the main challenges faced by OEMs in the transition to electric vehicles?

- How is the shift to electric vehicles affecting the automotive supply chain?

- What makes deep AI essential for solving complex industry challenges?

- What innovations in EV charging technology are expected to reduce range anxiety?

- How might bidirectional charging technology change the way we use electric vehicles?

So, the early adopters of these disruptors technicians, engineers, and software developers will need to stay ahead of innovations in autonomous systems, battery technologies, and connected vehicles to ensure all support of these technologies are effectively integrated into the workforce. In a major order the Governments, Industry Leaders, and educational systems must collaborate to ensure that the workforce is prepared for the future, in any formal or informal way. The urgency of the industry to re-skill the workforce is understood. A traditional education like a university career needs 2 to 5 years to get the degree. On the other side, digital educational platforms are emerging in dozens of miles and do not even guarantee the quality of skill and knowledge but allow cover the immediate requirement, a big risk to both [5, 6]. By focusing on education, training, and cross-sector collaboration, the electromobility industry can not only drive technological progress, but also create the New Jobs Roles (NJR), which will be well-paid jobs for the workers of tomorrow. In the end, it is perceived that the industry is not opening the benches for these new jobs, which the recruiters do not know how to make technical-knowledge-profiles that are required and end up just changing names to positions, doing what in Mexico we call rehashed-positions to cover the organizational chart.

2. Impact of EV technology adoption on jobs

As announced (WEF 2023) [1, 2], in the next 5 years, 83 million jobs are projected to be lost, and 69 million are projected to be created, constituting a structural labor-market churn of 152 million jobs. It is projected a reduction in employment of 14 million jobs. This sounds worrying, but it has several perspectives. The report published under the supervision of Saadia Zahidi, Managing Director of WEF, in the Future *of Jobs Survey 2023 and International Labour Organization, ILOSTAT*, shows that finally, the sum of these changes yields the estimated overall structural labor-market churn of 23% of the current global.

Now, let us take a look and make a close-up on the expected impact of technology adoption on new jobs in this overall structural labor market, which was estimated at a churn of 23%. The graph in **Figure 1** shows that all technologies, except two named

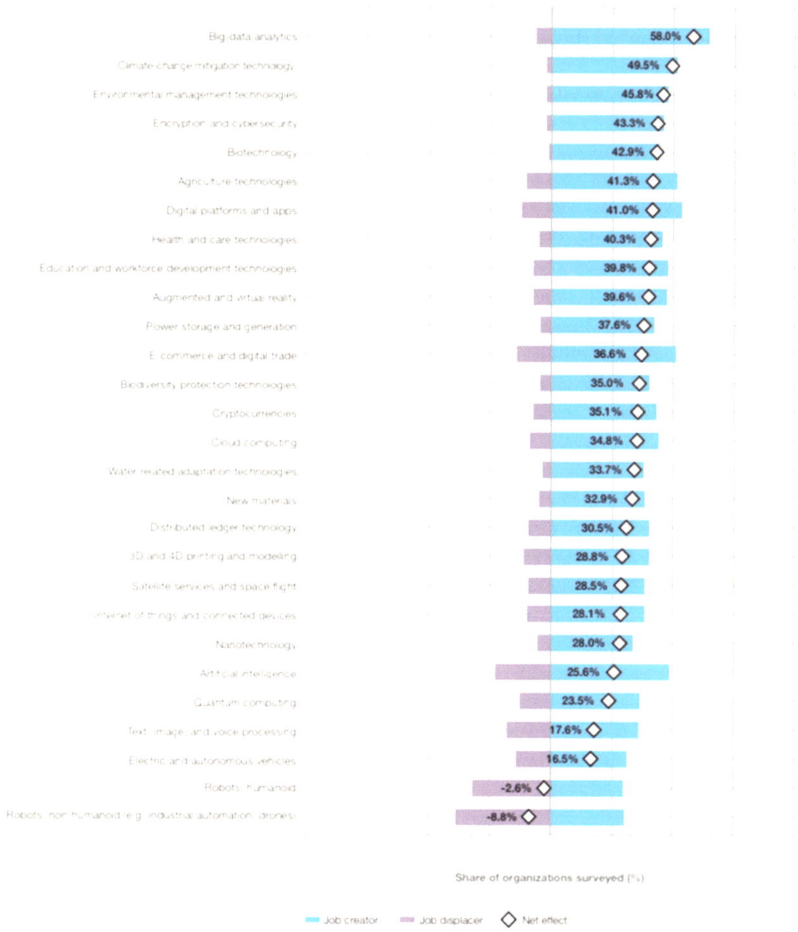

Figure 1.
Expected impact of technology adoption on jobs, 2023–2027 (Source: Future of Jobs Report 2023 WEF in Ref [1]).

"Robot-humanoid and Robot non-humanoid," are expected to create positive net jobs in the next 5 years (2030) [1, 7, 8]. Big Data, Data Analytics, Climate Change and Environmental Management Technologies, and Encryption and Cybersecurity are expected to be the main drivers of job growth. The net effect is positive, being greater in the creation of new jobs that will cover that churn in the current global.

Generative AI has received particular attention recently, with claims that 19% of the workforce could have more than 50% of their tasks automated by AI and job losses making headlines, while others expect the technology to improve jobs. According to recent data from the World Economic Forum, only robots-H and robots not-H are forecast to have an overall net negative impact on employment. Now, if we analyze the effect on labor supply and demand in the automotive sector due to the electromobility transition driven by climate change [9], the perspective as a real case does not look so encouraging. To do this, we will rely on an in-depth and very well-prepared study by the Automotive Skills Alliance (ASA) from 2021 to 2023 [10]. We will try to transfer the facts, and in the end, we will contribute with some important conclusions, crossed with events between the regions of North America (MX-USA-CN), LATAM, and isolated events of actions between the EU and the automotive industry of Asia.

2.1 Today, some concrete data on these changes in employment in the EV world

Europe news October 2024: Germany's auto industry prepares for job losses in the electric transition, presenting a collapse in its automotive industry. *Volkswagen is closing three plants in Germany, and Audi is halting production in Belgium, sparking fears of a crisis in Europe's car industry* [11].

Germany's auto industry is preparing for a big shift in employment as it transitions toward electric vehicles (EVs), while a recent study estimates it could lose over 186,000 jobs in the next decade by switching to EV production, which requires fewer employees than traditional car manufacturing. So, those potential losses, 46,000 jobs, disappeared.

Experts see this as an opportunity—if the sector can boost its competitiveness. "The attractiveness of locations for new investments and job creation will be decisive in the future," said Simon Schuetz, German Association of the Automotive Industry (VDA).

In North America, remember that on September 15, 2023, the United Auto Workers union (UAW), which represents around 150,000 American auto plant workers, declared a strike targeting three of the USA's automakers. The three manufacturers—General Motors, Ford Motor, and Stellantis. However, the parties agreed to lift the UAW strike on October 30, 2023. We can analyze some facts regarding employment related to the EV transition.

The headline of the WSWS on September 1, 2023, is surprising. "Stop the slaughter of jobs by electric vehicles". International Workers Alliance of Grassroots Committees.

At that time, one of the underlying reasons for the UAW was not only a wage adjustment, which was more lagging, but it is known that the UAW had an intelligence report on the impact in the United States on lost jobs during the transition process of EV manufacturing, which in summary expressed it would be devastating. Brett Smith, chief Technology Officer at the Center for Automotive Research (CAR), says: "The industry is going through an unprecedented transition" [12]. It is very likely that there will be fewer people making these cars and fewer people making the parts for these cars, and that will create problems in some car communities.

The predictions go from bad to worse. According to an "optimistic" study by the Economic Policy Institute (EPI 2023), in the next 7 years, 74,000 jobs could be eliminated in the automotive sector in the United States, of which 33,000 would be in assembly and 41,000 in auto parts. Finally, all this information has a social affectation and modifies the feelings of the massive number of buyers who rejected the acquisition of new EV cars, as was shown by Anmol Soni at Apr 22–2020 AYSPS Dissertations. Driving Green: Employment effects, policy adoption, and public perceptions of electric vehicles. Andrew Young School of Policy Studies.

Data on the automotive industry in Mexico in 2020, the automotive industry generated around 980,000 direct jobs and more than 3.5 million indirect jobs (Thompson Reuters Mexico) [13]. Actually, in 2024, there was reported 839,156 direct labor with trends growing 1.5% annually (Statista) [5], even though the net lost jobs with respect to 2020 forward is close to −14.38%.

Nearshoring is an investment opportunity for Mexico that can boost its economy and generate more employment. In 2023, foreign investment in Mexico related to nearshoring grew by 47% between January and September. Some relevant dates.

- Mexico is the largest automobile manufacturer in Latin America.
- Mexico is one of the main investment destinations for automotive companies.
- The Volkswagen plant in Puebla is the largest in Mexico.
- Mexican automotive industry represents 20.5% of the manufacturing GDP.
- Nearshoring has a positive impact on Mexico's economy; with the correct political climate, the investment will be close to $80.0 Billion USD.

2.1.1 Challenges facing the automotive sector in Mexico

- Persistence of the deficit in infrastructure and technology.
- Urgency to renew by re−/upskilling the base of automotive workers for the transition of EVs to almost 500 thousand people.
- Harsh water crisis in the north and center of the country.
- Security problems in key regions.

2.2 Outlook NJR and skill for EV 2023: 2030

Automotive Skills Alliance (ASA) in the European Union reports on a study focused on the development of skills for human resources and considers the entire value chain required for electromobility, of which the creation of more than 40 New Jobs Roles (NJR) for the Automotive EVs sector, and analyzes the new competencies that must be developed and promoted [10]. Also, launch a HUB Skill platform within an open collaboration to participate as providers of specialized training and training demands directed on specific projects based on these new work roles. ASA has promoted this initiative since 2018 (**Figure 2**), sustaining it for more than 5 years, with incredible evolution. https://automotive-skills-alliance.eu/downloads. It arises through Project

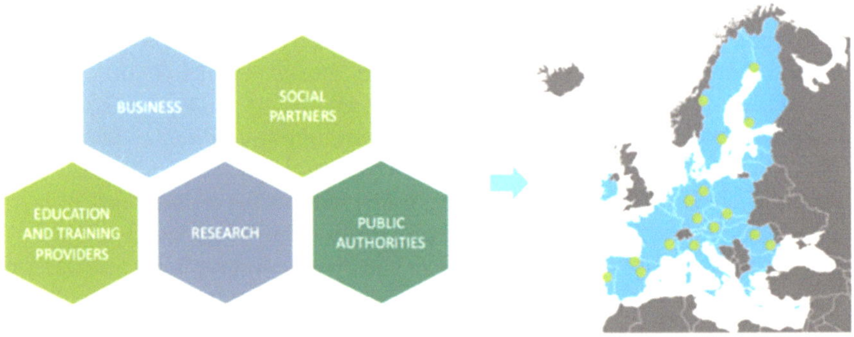

Figure 2.
Automotive Skills Alliance Community of European Union, including more than 11 countries. Source: ASA Office.

DRIVES (Development and Research on Innovative Vocational Skills)—Project number 591988-EPP-1-2017. The European Commission references [7, 8, 10].

2.3 The ambition

The aim is to put in place an up–/reskilling framework that maximizes automotive industry competitiveness, job retention, and new job opportunities—paving the way to a skills partnership for the whole automotive ecosystem.

It will present a roadmap toward the transformation of skills while also contributing to climate change resilience and other challenge recovery. It further broadens the outcomes of DRIVES (blueprint project), which will provide a key input to the work of the eMobility Skills Partnership.

The partnership also builds on inclusivity and openness for all relevant stakeholders, initiatives, or projects (e.g., ALBATTS, ECEPE, DRIVES blueprint projects) to jointly cooperate under one European umbrella [10]. The ambition of the partnership is to enable the up–/reskilling of 5% of the workforce each year to tackle the industrial transition in the automotive ecosystem. Of the 15 million workers in the European Automotive Industry, it is estimated that 10% are direct manufacturing and assembly workers who require re/up skilling in the transition to EV.

3. Normalizations of skill

Based on the study D2.9.1, preliminary strategic analysis DRIVES Project, with the ASA Office in particular of our colleague Dr. Jakub Stolfa, ASA President [7]. The skills have been identified from previous demand-survey normalization activity: 5 main categories or clusters, 4 of these being "technical" and the 5th related to previously identified "soft skills." Specific job roles comprise different combinations of these skills. Outlines the structure survey and its constituent parts. This is briefly in the following **Figure 3**.

3.1 EV job roles comprise different combinations of set skills (ASA)

1. *Technical knowledge profiles*: This category refers to the background of people in terms of education and expertise (at a high level (engineer) or medium level (technician)). It includes the following normalized skills (in alphabetical order):

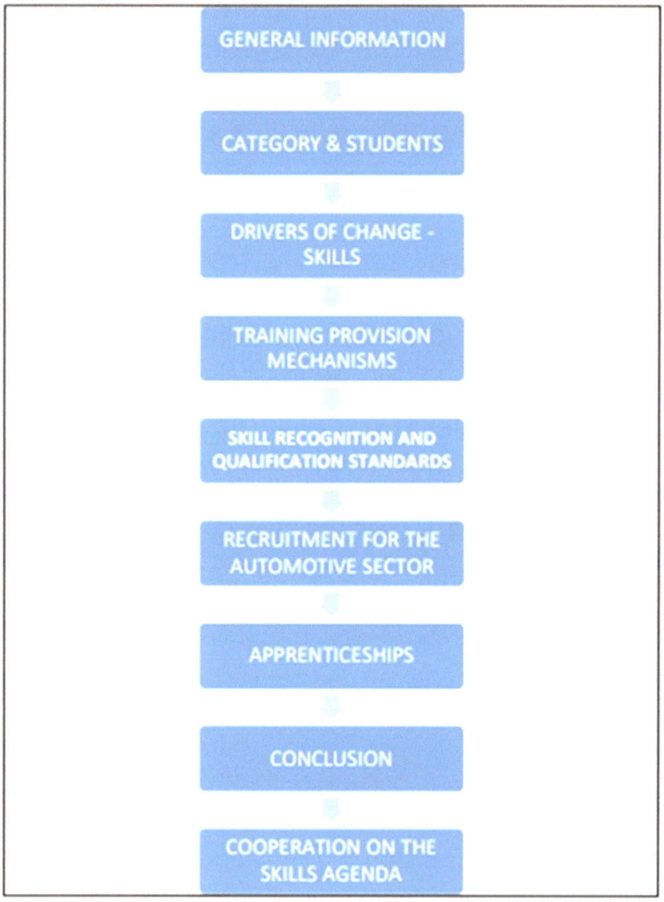

Figure 3.
Questionnaire structure (Source: ASA study D2.8).

Electrical; Electronic; Electrochemical; Material Sciences; Mechanical and Mechatronics; Software development; Sustainability; Technical knowledge (generic).

1. *Vehicle systems*: This category refers to knowledge and expertise relating to the different new systems and functions in the vehicle as a product (in alphabetical order). It includes the following normalized skills (in alphabetical order):

Alternative ICE powertrains (alternative fuels); Automated driving; Connectivity; Drivetrain Electrification (exist sub-categories); Functional safety; System architecture; Electrification Batteries; Electric motors; Energy management; Power electronics; System Integration; Thermal management.

2. *Life cycle/product*—process chain: This category refers to the different steps in the vehicle life cycle, covering the whole product and process chain (in sequential order). It includes the following normalized skills (in alphabetical order):

Market analysis; R&D&I; Automotive Design; Product Development; Simulation; Testing/Validation; Process engineering; Production/Manufacturing (see sub-categories below). Internal Logistics Sales After-sales service Mobility. Production/Manufacturing: Automation/Robotics, Maintenance. Production/ Organization, Specific Manufacturing Processes (Casting, Injection molding, Paint spraying, Precision machining.

3. *Digitalization*: This category refers to the specific digital enablers and the digital skills (in alphabetical order). It includes the following normalized skills (in alphabetical order):

3D printing; Artificial Intelligence; Big data/Data Analytics; Cybersecurity; Digital networks; Digital skills (generic); Digital twins IoT and Cloud; Predictive maintenance; Virtual product development and Virtual Testing.

4. *Soft skills* (in alphabetical order): adaptability/flexibility, behavioral agility, change management, communication, continuous improvement, creativity, critical thinking, entrepreneurship, foreign languages, learnability, management and leadership, networking, problem-solving, project management, resilience, teamwork.

3.2 Result of the survey

A total of 83 respondents completed or partially completed, the survey. Respondents are divided into the following categories for analysis [7, 8]:

- Accreditation, certification, or qualification body
- College
- Private company (excluding other categories listed)
- Research institute/center
- University
- VET school
- VET Umbrella organization

The next sheet shows the rankings of the skills list in **Tables 1–3** and the rankings of the job roles list in **Tables 4**–7.

3.3 Future job roles

It provides a solid multilayer base for automotive industry-specific New job roles and skill as showm in the **Figure 4**, how was reported for ASA in European-wide databases such as ESCO [7, 8, 10]. In other words, it serves as a version of the competence matrix for specific automotive industries. The goal is to become a reference definition resource that can be mapped or recognized at the stakeholder level. It also constitutes an aligned base through the Skills Hub by region (**Figure 4**).

% [Based on normalized skills (number of times the skill is nominated)]		
Ranking	**Skill name**	**%**
1	Big data/data analytics	6.8%
2	Software development	5.4%
3	Technical knowledge	4.1%
4	Learnability	3.7%
5	Mechatronics	3.4%
5	Materials sciences	3.4%
5	Digital skills	3.4%
5	Electrical/electronic	3.4%
6	System integration	3.1%
6	Adaptability/flexibility	3.1%
6	Cybersecurity	3.1%
7	Specific manufacturing processes	2.7%
7	Artificial intelligence	2.7%
7	R&D&I	2.7%
7	Market analysis	2.7%
8	Automation/robotics	2.4%
9	Management and leadership	2.0%
9	Process engineering	2.0%
9	Testing/validation	2.0%
9	Design	2.0%

Table 1.
Skill ranking 1/3.

Whether you are a company seeking trending courses for employees, an individual aiming to embark on a new learning journey, an education or training provider searching for job role-based curricula, or a regional/national representative seeking strategies, you can explore those new skill sets.

Based on the needs of the industry identified via Survey Demand, the DRIVES project partnership has a look at the set of job roles that the DRIVES project partnership is working on: https://www.project-drives.eu/en/driveslearningplatform [8].

- Each job role is publicly available.
- Short fact sheet.
- Skill cards and documents describing the learning outcomes.
- Online browsing possibility.

Figure 5 shows the relationship between skill index and skill ranking. The importance of demand ranking for the different sectors is evident, with big data/

% [Based on normalized skills (number of times the skill is nominated)]		
Ranking	**Skill name**	**%**
10	Automated driving	1.7%
10	Batteries	1.7%
10	System architecture	1.7%
10	Product development	1.7%
11	Communication	1.4%
11	Functional safety	1.4%
11	After-sales services	1.4%
11	Sustainability	1.4%
11	Energy management	1.4%
11	Creativity	1.4%
12	Mechanical	1.0%
12	Drivetrain	1.0%
12	Connectivity	1.0%
12	Change management	1.0%
12	3D printing	1.0%
12	Power electronics	1.0%
12	Project management	1.0%
12	Production organization	1.0%
12	Internal logistics	1.0%

Table 2.
Skill ranking 2/3.

% [Based on normalized skills (number of times the skill is nominated)]		
Ranking	**Skill name**	**%**
13	Thermal management	0.7%
13	Foreign languages	0.7%
13	Simulation	0.7%
13	Mobility services	0.7%
13	Teamwork	0.7%
13	Digital networks	0.7%
13	Virtual product development and testing	0.7%
13	Electric motors	0.7%
13	Sales	0.7%
13	Continuous improvement	0.7%
13	Problem solving	0.7%
14	Resilience	0.3%
14	IoT and cloud	0.3%

% [Based on normalized skills (number of times the skill is nominated)]		
Ranking	**Skill name**	**%**
14	Critical thinking	0.3%
14	Electrochemical	0.3%
14	Networking	0.3%
14	Digital twins	0.3%
14	Entrepreneurship	0.3%
14	Behavioral agility	0.3%
14	Alternative ice powertrains	0.3%
14	Maintenance	0.3%
14	Predictive maintenance	0.3%
14	Optimize activities	0.3%

Table 3.
Skill ranking 3/3.

% [Based on normalized skills (number of times the skill is nominated)]		
Ranking	**Job role name**	**%**
1	Automotive data analyst	6.4%
2	Powertrain engineer	4.9%
3	Automotive technician	4.4%
4	Innovation manager	3.9%
5	Cybersecurity engineer	3.4%
5	Design engineer	3.4%
5	Marketing and business development manager	3.4%
6	Automotive manager	2.9%
6	Software specialist	2.9%
6	Manufacturing engineer	2.9%
7	Electrification engineer	2.5%
7	Process engineer	2.5%
7	Systems engineer	2.5%
7	Maintenance technician	2.5%
8	Advanced material engineer	2.0%
8	Project manager	2.0%
8	Production worker	2.0%
8	Automotive engineer	2.0%
8	Software engineer	2.0%

Table 4.
Job roles ranking 1/4.

% [Based on normalized skills (number of times the skill is nominated)]		
Ranking	Job role name	%
9	Quality engineer	1.5%
9	Artificial intelligence professional	1.5%
9	Mobility services manager	1.5%
9	Architecture and integration specialist	1.5%
9	Sales specialist	1.5%
9	Battery professional	1.5%
9	Mechatronic engineer	1.5%
9	Digitalization engineer	1.5%
10	Product engineer	1.0%
10	Mechanical technician	1.0%
10	Vehicle data specialist	1.0%
10	3D printing technician	1.0%
10	Customer care assistant	1.0%
10	Compliance manager	1.0%
10	Electrical technician	1.0%
10	V2X specialist	1.0%
10	Quality technician	1.0%
10	Product manager	1.0%
10	Mechatronic technician	1.0%

Table 5.
Job roles ranking 2/4.

% [Based on normalized skills (number of times the skill is nominated)]		
Ranking	Job role name	%
11	Homologation engineer	0.5%
11	Mechanical engineer	0.5%
11	Technical sales support	0.5%
11	Maintenance engineer	0.5%
11	System integration engineer	0.5%
11	Accessibility manager	0.5%
11	Chief communications officer	0.5%
11	Diversity manager	0.5%
11	Sustainability manager	0.5%
11	Technical writer	0.5%
11	Terminology manager	0.5%
11	Configuration and change manager	0.5%
11	Autonomous driving specialist	0.5%

% [Based on normalized skills (number of times the skill is nominated)]		
Ranking	**Job role name**	**%**
11	Computer vision engineer	0.5%
11	Fleet management system developer	0.5%
11	Automation engineer	0.5%
11	Logistics manager	0.5%
11	Partner relationship manager	0.5%

Table 6.
Job roles ranking 3/4.

% [Based on normalized skills (number of times the skill is nominated)]		
Ranking	**Job role name**	**%**
11	Aces engineer	0.5%
11	Dependability engineer	0.5%
11	Power electronics engineer	0.5%
11	Head of strategy	0.5%
11	Human resources specialist	0.5%
11	Procurement professional	0.5%
11	Robotic engineer	0.5%
11	Rubber technologist	0.5%
11	Alternative energy engineer	0.5%
11	Sales manager	0.5%
11	Additive manufacturing engineer	0.5%
11	R&D engineer	0.5%
11	Cybersecurity manager	0.5%
11	Application engineer	0.5%
11	Industrial engineer	0.5%
11	Technical manager	0.5%
11	Electrical engineer	0.5%
11	Production team leader	0.5%
11	Government affairs	0.5%
11	Capacity planner	0.5%
11	Key account e-mobility	0.5%

Table 7.
Job roles ranking 4/4.

data analytics being the most relevant, even above AI. **Figure 6** shows an important difference between the top 10 demand and the current educational offer, evidencing an outdatedness of the educational sector since only 5 top normalized skills can be detected in 20 skills offered.

Figure 4.
Future job roles list for the EV automotive industry.

3.4 New roles and opportunities SUMAN, PM Ing and manufacturing engineering (NJR)

The purpose of the Set Skill Card is to define the details of "re–/up skills" for 3 NJRs (**Figures 7** and **8**). There, we shared the Set Skill Card of Sustainability Manager (SUMAN) and Predictive Maintenance Eng (PM-Eng). While ML-Eng is the third NJR of **Table 1** within the ECQA skills definition model. *(ECQA standards are used to describe the skill sets delivered within the DRIVES project) (www.projectdrives.eu)* [7, 8].

The rapid growth of autonomous driving technologies, particularly within the electric vehicle (EV) sector, is transforming the automotive industry. Central to this

DOI: http://dx.doi.org/10.5772/intechopen.1009441

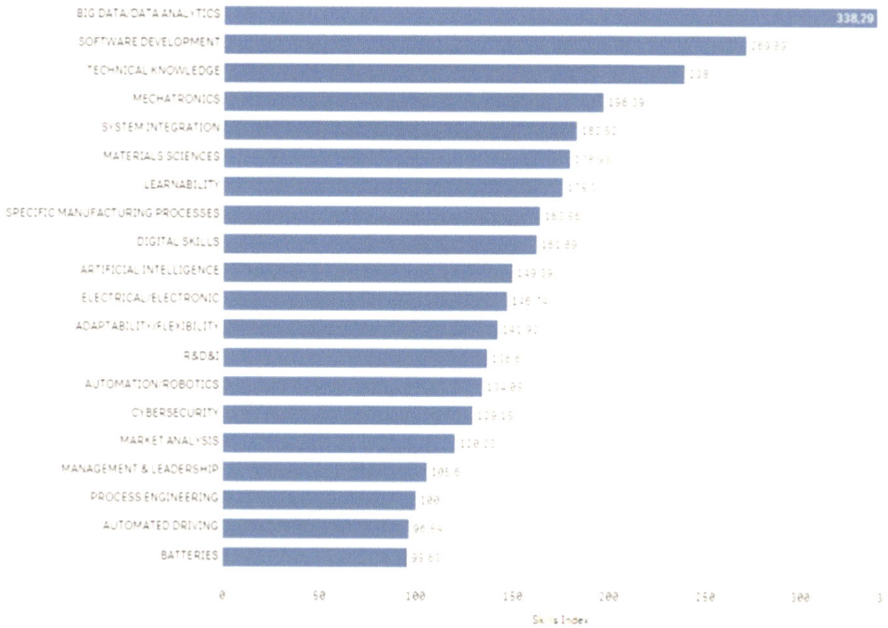

Figure 5.
Skill Index (demand) - overall sample.

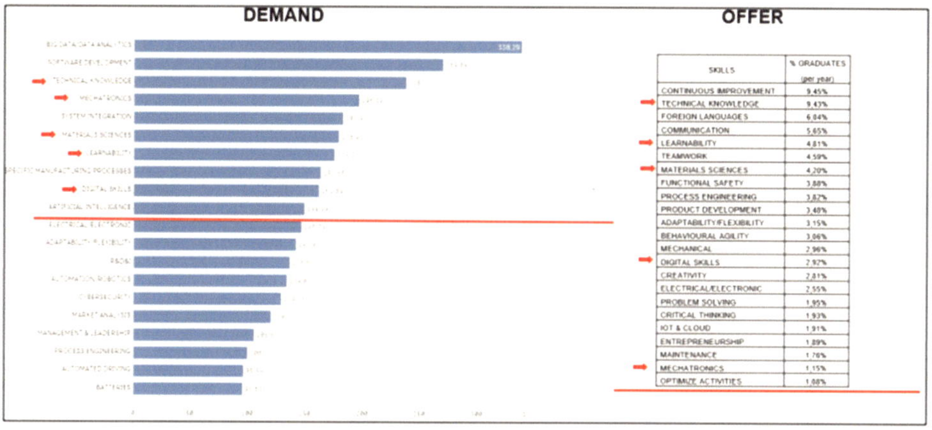

Figure 6.
Difference between demand and offer.

transformation is the application of machine learning (ML), which enables vehicles to "learn" from data and experience belonging to the data science, allowing them to make decisions like predictive maintenance engineers or generative design without being explicitly programmed.

Machine learning allows for adaptive, real-time decision-making in autonomous EVs, overcoming the limitations of traditional rule-based programming. Instead of pre-programming responses for every possible situation, autonomous systems can process vast amounts of data (e.g., sensor inputs, traffic patterns) to continuously improve driving behavior. This is particularly important as vehicles

Figure 7.
Skill Card of the Sustainability Manager (SUMAN).

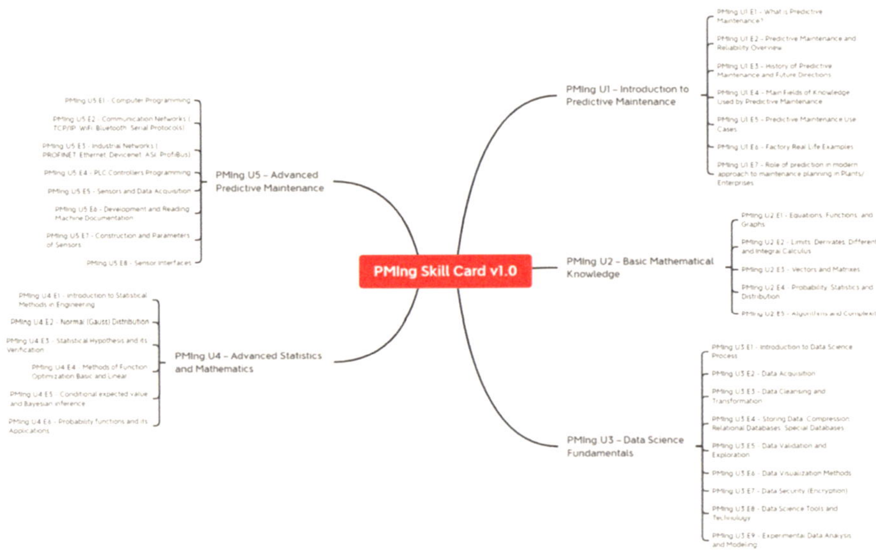

Figure 8.
Skill Card of Predictive Maintenance Engineer link to data science ML/AI (PM-Eng).

become more complex, integrating technologies like *LIDAR*, *radar*, *cameras*, and *advanced onboard computing*.

As the industry shifts to *self-driving electric vehicles*, this creates a demand for new job roles (see Refs. [14–16]):

AI/ML engineers: Developing and optimizing algorithms for autonomous systems.

Autonomous vehicle testers: Testing real-world performance and ensuring safety.

Cybersecurity experts: Protecting vehicles from data breaches and attacks.

Human-machine interaction specialists: Ensuring intuitive user interfaces for autonomous driving.

Software engineers and data scientists: Working on the integration of AI and vehicle systems.

Manufacturers will also need to adapt production lines for the integration of complex AI and sensor technologies, creating opportunities for *software developers*, *system integrators*, and *safety engineers*.

The rapid growth of "self-driving technologies"—especially as they relate to electric vehicles (EVs)—is one of the most exciting developments in the "electromobility sector." As you rightly pointed out, the integration of "machine learning (ML)" into vehicle platforms, particularly for autonomous driving, has huge implications not only for the technology itself but also for the workforce that will need to develop, operate, and maintain these systems.

4. Aspect on set skilling for machine learning (ML) and autonomous electric vehicles

4.1 The emergence of autonomous electric vehicles

- *The rise of self-driving EVs*:

  - Highlight the growing trend of integrating autonomous driving features with electric vehicle platforms. Companies like *Tesla*, *Waymo*, *Cruise*, and *Baidu* are leading the way, demonstrating how self-driving capabilities are rapidly evolving in tandem with electric mobility.

  - EVs are especially well-suited for autonomous driving due to their software-driven nature and the shift from traditional mechanical systems to more electrical, computerized systems. This creates synergies between electrification and autonomous driving that traditional vehicles could not easily achieve.

- *Technological advancements in machine learning*:

  - *Machine Learning (ML)*, especially *deep learning* and *reinforcement learning*, is transforming the way vehicles are able to perceive the environment, make decisions, and operate autonomously.

  - These technologies enable the vehicle to continuously learn from real-world driving data, improve its performance, and adapt to new situations without needing to be explicitly programmed.

4.2 Machine learning and the complexity of autonomous systems

- *Complexity of autonomous driving systems*:

  - As vehicles become increasingly complex, it is nearly impossible for manufacturers to create every rule or scenario for how the system should respond in every possible situation. Traditional programming models cannot account for the unpredictable nature of driving, from weather conditions to human behaviors and interactions.

- ML allows for *adaptive decision-making*: Instead of programming a specific response to every potential situation, ML systems allow vehicles to make real-time decisions based on data patterns learned from experience. The vehicle learns from vast amounts of data—such as video feeds, radar, lidar, and GPS—to interpret and act within its environment.

- *End-to-end learning systems*:
  - Many self-driving platforms, particularly those developed by tech companies, use *end-to-end learning* approaches, where a deep neural network processes raw sensor data (e.g., camera images, radar scans) to directly predict the vehicle's actions (e.g., steering, acceleration, braking).
  - These systems are designed to become *increasingly autonomous*, adapting to new road conditions and scenarios over time. This ability to "learn on the job" offers a unique advantage compared to traditional rule-based systems.

4.3 Workforce transformation: New roles and skill sets in autonomous EV manufacturing and development

- *Upskilling sets for autonomous EV platforms*:
  - *Data scientists and ML engineers*: As vehicle manufacturers and tech companies invest heavily in AI and ML, there's a growing demand for experts who can design, train, and optimize these machine learning models. Engineers with expertise in *artificial intelligence*, *machine vision*, and *natural language processing* will play crucial roles in developing these systems.
  - *Autonomous systems testers*: While machine learning models can improve over time, they need to be rigorously tested to ensure they are safe and reliable. This involves *virtual simulations*, *closed-course testing*, and *real-world driving trials*. Technicians will need specialized knowledge of testing algorithms and autonomous systems.
  - *Cybersecurity professionals*: With increased connectivity and reliance on software and data, securing autonomous systems is paramount. There will be a growing need for cybersecurity professionals who can protect against threats to vehicle autonomy, data integrity, and system hacking.
  - *Human-machine interaction specialists*: As vehicles interact more with human drivers and pedestrians, designing interfaces that are intuitive and safe will become a priority. Experts in *human factors engineering* and *user interface (UI) design* will be in demand to ensure the transition to autonomous driving is seamless.

- *Reskilling traditional automotive workers*:
  - Just as with other areas of EV manufacturing, workers with a background in traditional automotive engineering may need to re-skill in areas like AI, software development, and data analytics.

- For example, *mechanical engineers* transitioning into autonomous vehicle development might need to focus more on *systems engineering*, *robotics*, and *embedded systems programming*.

- Manufacturers will likely partner with universities and online platforms to offer specialized courses that blend automotive knowledge with cutting-edge skills in AI, ML, and autonomous systems.

4.4 The impact of autonomous driving on manufacturing and vehicle production

- *Redesigning production lines for autonomous EVs*:

    - Autonomous EVs require entirely different production lines compared to traditional vehicles. For instance, components like *LIDAR* sensors, *radar*, *cameras*, and advanced *onboard computing systems* need to be integrated with the EV's existing electrical infrastructure. Manufacturing processes may need to adapt to incorporate these high-tech sensors and computing platforms into the assembly line.

    - The increased role of software also implies that *software engineers* and *system integrators* will play a bigger role in the production process. Companies like *Tesla* are already using *over-the-air (OTA) software updates* to refine and improve autonomous features after the car has been sold, making the manufacturing process much more intertwined with software development.

- *Testing and safety protocols*:

    - The production of autonomous vehicles will demand stricter safety protocols. The technology needs to be rigorously tested, not only in simulations but also in real-world environments, including complex urban traffic, extreme weather conditions, and edge-case scenarios. This shift will create new roles for *autonomous vehicle safety engineers* and *regulatory compliance experts*.

    - There will also be new challenges in ensuring that autonomous vehicles can co-exist safely with non-autonomous vehicles. As automation becomes more common, workers may need to focus on *V2X (Vehicle-to-Everything) communications*, improving interaction between self-driving cars, other vehicles, and infrastructure.

4.5 Challenges and ethical considerations in autonomous driving and employment

- *Ethical dilemmas in autonomous systems*:

    - As machines begin to make decisions about safety, ethics becomes a key issue. Who is responsible if an autonomous vehicle is involved in an accident? How do we ensure that the algorithms make the right decisions in life-or-death situations? These are pressing questions that will shape both the development of autonomous EVs and the regulation of these technologies.

  - This will create demand for professionals in *ethics and AI governance*, helping manufacturers navigate the moral and regulatory challenges of autonomous driving.

- *Impact on the workforce*:

  - The rise of *autonomous driving* could eventually reduce the demand for certain jobs, especially in driving-based industries (e.g., truck drivers and delivery drivers). However, as with other technological advancements, new jobs will be created in the areas of AI development, safety regulation, and systems testing.

  - Job displacement may be a concern, but it is also an opportunity for upskilling and reskilling workers, especially in industries connected to logistics, public transportation, and maintenance.

  - The need for reskilling and upskilling.

- *The skills gap*: Many workers in traditional automotive manufacturing may not have the skills needed for EV production. For example, mechanics who have worked on ICE engines will need new training to work on electric drivetrains.

- *Key skills for the future*:

  Electric vehicle technology: Understanding battery chemistry, electric motor systems, and charging technology.

  Advanced manufacturing techniques: Familiarity with automation, robotics, and digital manufacturing tools, which are increasingly used in EV production.

  Software and data analytics: The integration of advanced software for vehicle performance, energy management, and driver assistance technologies.

  Sustainability knowledge: Awareness of green manufacturing processes and sustainable supply chains.

- Training and development programs

  Collaboration with educational institutions: Highlight how manufacturers and governments are partnering with universities, trade schools, and online platforms to offer specialized training programs.

  Vocational training: Programs that offer hands-on experience with EV assembly, battery technology, and charging infrastructure.

  Certification programs: Industry-recognized certifications for EV mechanics, battery technicians, and engineers.

  Apprenticeships and on-the-job training: Work-based learning opportunities within established automotive companies or new startups.

Online learning platforms: Growth of online courses (e.g., Coursera, edX) offering access to technical skills related to EVs and electromobility.

4.6 Challenges and barriers to workforce transformation

Resistance to change: Some workers might be resistant to retraining, especially if it requires a significant change in career trajectory.

Geographic disparities: The transition to EV manufacturing may be uneven across regions, with some areas struggling to provide sufficient training and job opportunities.

Time and cost of training: Training programs can be expensive and time-consuming, which may deter some workers from pursuing new opportunities, especially if they are already established in their current roles.

Rural and legacy manufacturing hubs: Areas with a long history of internal combustion engine vehicle manufacturing may face a more difficult transition due to the existing workforce's skill sets and the higher cost of retraining.

5. Case studies and best practices

Tesla's Gigafactory and workforce training: Look at how Tesla has approached workforce development and how it has influenced the broader EV industry. Tesla's Employee Training Evolution. https://www.rallyware.com/blog/training-and-auto-technology-evolving-together-in-the-industry [14, 17].

Germany's transition to EV manufacturing: Discuss Germany's "Industrie 4.0" initiative and its focus on reskilling workers for digital manufacturing, automation, and EV production [15, 17].

China's rapid EV expansion: Explore China's approach to scaling EV manufacturing while simultaneously addressing workforce needs, especially in cities like Shenzhen [16, 18].

6. Conclusion: Preparing for the future of autonomous electric vehicles

In particular, building the workforce of the future

- *The future of jobs in electromobility*: Summarize the long-term impact of the transition to EVs on job creation and the transformation of the workforce.

- *Call to action*: Stress the importance of continued investment in workforce development, cross-industry collaboration, and government policy to ensure a smooth transition.

- *Data science*: The integration of *machine learning* with electric mobility will revolutionize the automotive industry, creating new career opportunities and challenges. As the technology matures, the workforce will need to continually adapt, acquiring new skills in software development, AI, and systems engineering.

- *Education:* It is crucial that the industry continues to invest in education and training, both to maintain a skilled workforce and to ensure that these new technologies are implemented safely and ethically. Governments, companies, and educational institutions must collaborate to create a roadmap that addresses both the technical and human aspects of this transformation.

This section will help you bridge the gap between *autonomous driving technology* and the future workforce in the electromobility industry. It not only ties in with your focus on workforce development and re–/upskilling but also explores how the advanced technologies of AI and machine learning are shaping the jobs of tomorrow.

The transition to electric vehicles cannot be allowed to come at the expense of workers. Social ownership and democratic control of the automotive industry by workers is necessary so that this transition can be carried out through a massive improvement in workers' living standards and working conditions around the world.

In capitalism, advances in artificial intelligence, robotics, and other labor-saving technologies are used to throw workers into unemployment. But if controlled by the workers themselves and managed in accordance with their interests, advances in humanity's productive capacity can be used to reduce the physical wear and tear of workers, shorten the work week, and ensure a high standard of living for all. The change could be easier.

Acknowledgements

We would like to thank the great ASA Office community, particularly Dr. Jakub Stolfa, President of the ASA, and Mr. Petr Dolejsi, Vice-president of ASA (Europe Union). Also, to our colleagues at CIAC (Automotive Cluster of Coahuila, Mexico). Finally, support to the *DATASc CG office in Mexico and DATASc Dallas Fort Worth, Tx. Offices, USA*.

Author details

Jesus M. Orona-Hinojos*, Alejandro Yamaguchi Torres-Valdez,
Lizbeth A. Huerta-Larrumbe, Eduardo Macias-Avila and Alfredo Rios-Nogues
DataSc Consultancy Group, Saltillo, Coahuila, Mexico

*Address all correspondence to: jesus.orona@datasc-consultant.com

IntechOpen

© 2025 The Author(s). Licensee IntechOpen. This chapter is distributed under the terms of the Creative Commons Attribution License (http://creativecommons.org/licenses/by/4.0), which permits unrestricted use, distribution, and reproduction in any medium, provided the original work is properly cited.

References

[1] World Economic Forum. The Future of the Jobs Reports. World Economic Forum; 2023. Available from: https://www.weforum.org/publications/the-future-of-jobs-report-2023/

[2] Federal Media Network, LLC. Green Jobs Transportation. (s.f.). UK, Y. E. 2022. Available from: https://ehsjobs.org/greenjobs-transportation/

[3] Green Careers: Explore Green Tech Trends Driving the Automotive Industry. School of Thought. Youth Employment UK. . Available from: https://www.youthemployment.org.uk/green-careers-explore-green-tech-trends-driving-the-automotive-industry-school-of-thought/

[4] Green Jobs in India - Blog - Institute for Sustainable Communities. Institute For Sustainable Communities. Available from: https://sustain.org/blog/green-jobs-in-india/

[5] Spaeth K, Björnbäck W. Talent retention: Retaining young employees in a large and established organization with an aging workforce - a case study within Volvo group digital & IT [thesis]. Sweden: University of Gothenburg; 2023

[6] Descos L, Gospic A-M. Attract and Retain Generation Z in the White-Collar Business Sector. Sweden: Jönkönping University, Jönkönping International Business School; 2019. Available from: https://www.diva-portal.org/smash/get/diva2:1320600/FULLTEXT01.pdf

[7] Project DRIVES. 2021. Available from: https://www.project-drives.eu/en/results

[8] Project DRIVES. 2023. Available from: https://www.project-drives.eu/Media/Pages/30/01_Job_Roles_List.pdf

[9] Strategy & Part of the PwC Network. Transforming Vehicle Production by 2030 and How shared mobility and automation will revolutionize the auto industry. 2018. Available from: https://www.pwc.com/hu/hu/kiadvanyok/assets/pdf/transforming_vehicle_production.pdf

[10] The ECQA Certified Electric Powertrain Engineer project (ECEPE), ECQA Certified Electric Powertrain Engineer. Available from: https://conference.eurospi.net/images/eurospi/2021/ECEPE-Powertrain-Engineer-TRAINING-SCHEDULE-Plan-2.0.pdf

[11] Germany's Auto Industry. Available from: https://www.euronews.com/embed/2665956

[12] Isidore C. Electric Vehicles Pose a Major Threat to Autoworkers' Jobs, CNN Business. 2019. Available from: https://www.cnn.com/2019/12/04/business/electric-car-job-threat/index.html

[13] Industria automotriz en México: avances y amenazas en 2024. 2024. Available from: https://www.thomsonreutersmexico.com/es-mx/soluciones-de-comercio-exterior/blog-comercio-exterior/industria-automotriz-en-mexico

[14] Tesla's Employee Training Evolution. Available from: https://www.rallyware.com/blog/training-and-auto-technology-evolving-together-in-the-industry

[15] Xun X, Yuqian L, Vogel-Heuser B, Wang L. Industry 4.0 and industry 5.0—Inception, conception and perception. Journal of Manufacturing Systems. 2021;**61**:530-535. DOI: 10.1016/j.jmsy.2021.10.006

[16] Huld A. China's Industry Clusters – A Comprehensive Overview. 2024. Available from: https://www.china-briefing.com/news/chinas-industry-clusters-comprehensive-overview/

[17] Conzade J, Cornet A, Hertzke P, Russell H. Why the automotive future is electric. 7 Sept 2021. Available from: https://www.mckinsey.com/industries/automotive-and-assembly/our-insights/why-the-automotive-future-is-electric

[18] Robinson R, Reddy-Tummalapalli S. Climbing the electric vehicle transformation mountain. In: Deloitte C editor. Industry Center. 7 Sept 2023. Available from: https://www2.deloitte.com/us/en/insights/industry/retail-distribution/consumer-behavior-trends-state-of-the-consumer-tracker/ev-electric-vehicle-transformation

Chapter 11

Recent Advancements in Automotive Engineering by Using Evolutionary Algorithms and Nature-Inspired Heuristic Optimization

Morteza Mollajafari and Salman Ebrahimi-Nejad

Abstract

The integration of evolutionary algorithms and nature-inspired heuristic optimization has marked a significant advancement in automotive engineering. These methods, drawing inspiration from biological processes, have been instrumental in optimizing complex engineering problems, leading to more efficient, reliable, and high-performing automotive designs. The application of such algorithms has been particularly transformative in areas such as vehicle routing, predictive maintenance, and design optimization. These advancements not only signify a leap in the computational capabilities within the industry but also pave the way for the development of autonomous vehicles and smart transportation systems. The future of automotive engineering is poised to be heavily influenced by the continued evolution and application of these sophisticated algorithms, which promise to bring about even more groundbreaking innovations in the field. The potential for these technologies to revolutionize the automotive industry is immense, as they offer solutions to some of the most pressing challenges faced by engineers today. As these algorithms evolve, they will undoubtedly unlock new possibilities in vehicle efficiency, safety, and performance, marking a new era in automotive engineering.

Keywords: heuristic optimization, vehicle aerodynamics, automotive engineering, evolutionary algorithms, autonomous vehicles

1. Introduction

Evolutionary algorithms (EAs) are Artificial Intelligence (AI) methods inspired by natural selection, used to solve optimization and search problems by refining candidate solutions based on a fitness function to produce better offspring in subsequent generations [1, 2].

Evolutionary Algorithms (EAs) offer significant benefits in automotive engineering by optimizing complex and multi-objective systems. Their key advantages include multi-objective optimization, enabling the balancing of competing requirements like fuel efficiency, performance, and safety. Capability to manage non-linear systems without needing exact mathematical models, making them ideal for tasks like powertrain, aerodynamics, and suspension design.

Global search function, which helps avoid getting trapped in local optima when dealing with intricate design spaces. High flexibility and robustness [3, 4], which make EAs well-suited for working with uncertain and noisy data while adapting to evolving design needs. Applications across design and manufacturing*, including structural optimization, powertrain tuning, and the development of autonomous systems. EAs are particularly useful for optimizing aerodynamic designs, fine-tuning suspension systems, designing hybrid powertrains, and ensuring durability and reliability in automotive components.

While Evolutionary Algorithms offer several benefits in the automotive industry, they also come with certain limitations, including High Computational Cost, Scalability Issues, Parameter Sensitivity, Convergence to Suboptimal Solutions [5–7], Integration Challenges in Real-Time Systems, Limited Analytical Insight, and Need for Customization. These limitations suggest the need for hybrid approaches to improve effectiveness and adaptability in automotive engineering.

Heuristic optimization plays a pivotal role in automotive engineering, serving as a cornerstone for innovation and efficiency in the design and manufacturing processes [8]. Heuristic methods, inspired by natural processes, help engineers tackle complex challenges by finding optimal solutions where traditional methods fail. They are valuable in applications like vehicle weight reduction, improving fuel efficiency, and lowering emissions [9]. Moreover, heuristic optimization extends its benefits to the realm of safety and durability [10].

Integrating heuristic optimization with machine learning and high-performance computing boosts its effectiveness and impact [11].

In **Table 1**, systematic comparison between the various algorithms is made.

These disadvantages highlight the trade-offs involved in selecting the appropriate algorithm for specific applications in automotive engineering.

While significant advancements have been made in the application of evolutionary algorithms and nature-inspired heuristic optimization in automotive engineering, several critical gaps remain. This chapter aims to address these gaps by focusing on the following areas:

1.1 Limited integration of hybrid approaches

Although individual algorithms such as Genetic Algorithms (GAs), Particle Swarm Optimization (PSO), and Ant Colony Optimization (ACO) have been widely applied, there is a lack of comprehensive studies on the integration of hybrid models. These hybrid approaches, which combine the strengths of multiple algorithms, are underexplored in their ability to solve highly complex, multi-objective optimization problems in automotive engineering [12].

1.2 Inadequate real-time optimization in autonomous systems

Current research has made progress in applying optimization algorithms to vehicle design and manufacturing, but there is a noticeable gap in the real-time application of

Algorithm	Inspiration	Strengths	Weaknesses	Applications in automotive engineering
Genetic Algorithm (GA)	Natural selection and genetics	Good for complex, multi-objective optimization; handles noisy and changing environments well	Can be computationally expensive; it requires careful parameter tuning	Aerodynamic optimization, powertrain design, suspension tuning
Evolutionary Strategies (ES)	Natural evolution	Adaptive mutation; strong in continuous optimization problems	May struggle with discrete problems or require large populations	Vehicle shape optimization, powertrain efficiency
Genetic Programming (GP)	Evolution of computer programs	Develops adaptable programs that evolve to solve problems	Prone to "bloat" (large, inefficient solutions); computationally intensive	Control systems optimization, autonomous vehicle navigation
Differential Evolution (DE)	Differential variation	Simple and effective; suitable for non-differentiable problems	May converge slowly on difficult optimization problems	Engine tuning, battery management system optimization
Particle Swarm Optimization (PSO)	Swarm behavior of birds/fish	Fast convergence; simple to implement	Can get trapped in local optima; may require manual tuning	Route optimization, engine parameter tuning
Ant Colony Optimization (ACO)	Ant foraging behavior	Good for path-finding and combinatorial optimization	May require significant computational resources, slower convergence	Vehicle routing, scheduling problems in manufacturing
Simulated Annealing (SA)	Physical annealing in metallurgy	Avoids local optima by allowing worse solutions initially	Convergence can be slow, requires careful cooling schedule	Manufacturing process optimization, engine calibration
Artificial Bee Colony (ABC)	Foraging behavior of honeybees	Efficient in exploring diverse solution spaces	May have slower convergence in some complex problems	Supply chain optimization, material selection
Algorithm	Inspiration	Strengths	Weaknesses	Applications in Automotive Engineering
Genetic Algorithm (GA)	Natural selection and genetics	Good for complex, multi-objective optimization; handles noisy and changing environments well	Can be computationally expensive; requires careful parameter tuning	Aerodynamic optimization, powertrain design, suspension tuning

Table 1.
Overview of optimization algorithms: inspiration, strengths, weaknesses, and applications in automotive engineering.

these algorithms in autonomous vehicle systems. Specifically, the challenge of optimizing navigation, decision-making, and safety in real-time scenarios with constantly changing variables has not been fully addressed [13].

1.3 Challenges in sustainable material selection

Research on material optimization for lightweight design has largely focused on traditional materials like carbon fiber composites and aluminum alloys. However, there is insufficient exploration into the use of sustainable and eco-friendly materials. The lack of research in this area presents a significant opportunity for developing more sustainable solutions for future vehicle designs [14].

1.4 Underexplored impact of machine learning on heuristic optimization*

While machine learning has been incorporated into some heuristic optimization methods, the full potential of integrating advanced AI techniques such as deep learning with evolutionary algorithms remains untapped [15]. The chapter will explore how this integration can enhance the adaptability and efficiency of optimization processes in automotive engineering.

By addressing these gaps, the chapter aims to contribute new insights and methodologies that will further advance the application of optimization algorithms in automotive engineering, particularly in the areas of hybrid algorithm development, real-time optimization, sustainable material selection, and AI integration.

2. Fundamentals of evolutionary algorithms

Before delving into specific applications, it is crucial to understand the core principles and methodologies that underpin evolutionary algorithms and heuristic optimization techniques. This section provides an overview of key algorithms and their unique characteristics.

1. Population: A set of potential solutions to the optimization problem.

2. Fitness function: Evaluates the quality of each solution.

3. Selection: Choosing the best-performing individuals for reproduction.

4. Crossover: Combining parts of two individuals to create offspring.

5. Mutation: Introducing random changes to maintain diversity.

EAs are particularly effective for solving complex problems with large search spaces or when traditional optimization methods struggle [16].

2.1 Genetic algorithms

Genetic Algorithms (GAs) are a fascinating and powerful subset of Evolutionary Algorithms (EAs), which are computational models inspired by the process of natural selection that occurs in biological evolution [17]. GAs have been successfully applied to a wide range of complex problems where traditional optimization techniques may struggle [18]. This section explains how GAs evolve solutions over time.

Core Concepts:

- *Population*: A set of potential solutions to the problem.
- *Individual*: A single solution within the population.
- *Chromosome*: The genetic representation of an individual.
- *Gene*: A part of a chromosome representing a specific parameter.
- *Fitness Function*: Evaluates how well an individual solves the problem.

Evolution Process:

1. *Initialization:* Generate an initial population randomly.
2. *Evaluation*: Assess each individual's fitness.
3. *Selection*: Choose individuals for reproduction based on fitness. Methods: Roulette Wheel, Tournament, or Rank Selection.
4. *Reproduction*:

   a. *Crossover*: Combine genetic information from parents. Types: Single-Point, Two-Point, or Uniform Crossover.

   b. *Mutation*: Randomly alter genes to maintain diversity. Types: Bit Flip, Gaussian, or Swap Mutation.

5. *Replacement*: Form a new generation, optionally using elitism to preserve top solutions.
6. *Iteration*: Repeat steps 2–5 for multiple generations
7. *Termination*: Stop when meeting criteria like generation count or fitness threshold.

Fitness function selection: The selection of an appropriate fitness function is essential as it directly affects the performance of a Genetic Algorithm. This function assesses how optimal a solution is, guiding the evolutionary process. While simple problems may require straightforward fitness functions, complex problems often involve multiple variables. For multi-objective optimization, Pareto-based functions are commonly used to manage trade-offs between competing objectives, such as fuel efficiency and engine power in automotive engineering. In such cases, a Pareto front is generated, representing a set of optimal solutions where no objective can be improved without negatively impacting another. Parameter Tuning, the success of a GA is heavily dependent on effective parameter tuning, which includes adjusting population size, mutation rates, and crossover probabilities. If the mutation rate is too high, the algorithm may behave too randomly, while too low a rate can cause premature convergence. Similarly, the size of the population influences how extensively the solution space is explored. These parameters must be carefully balanced to maintain diversity while still converging toward the best solutions. Adaptive GAs, which adjust parameters dynamically during the process, help optimize the balance between exploration and exploitation. Additionally, meta-GAs have gained traction as a means of automating the tuning of GA parameters. Handling Multi-Objective Optimization, When dealing with multiple, often conflicting, objectives, GAs require specialized methods to find optimal solutions across several dimensions. Multi-objective GAs, such as NSGA-II, are designed to preserve diversity within the population while simultaneously generating solutions that balance various objectives. These algorithms excel in practical scenarios, such as automotive design, where factors like safety, performance, and cost must all be optimized. The primary challenge in multi-objective optimization is ensuring both convergence to the Pareto front and maintaining diversity among solutions [2].

2.2 Evolutionary strategies

Evolutionary Strategies (ES) are a cornerstone of Evolutionary Algorithms (EA), a class of algorithms inspired by the process of natural selection. These strategies are particularly adept at solving optimization problems where traditional methods may falter, especially in complex, multimodal, or noisy search spaces. ES operates on the principle of "survival of the fittest," where a population of candidate solutions evolves over generations toward an optimal or satisfactory solution.

The fundamentals of ES are rooted in the concept of variation and selection. Variation is introduced through operators such as mutation and recombination, while selection is based on the fitness of the individuals within the population. This iterative process of generating variability and selecting the best candidates ensures that the population converges toward peak performance over time.

One of the key aspects of ES is its use of self-adaptation [19]. This means that the strategy parameters, which control the variation operators, are themselves subject to evolution. This allows ES to dynamically adjust its behavior in response to the changing landscape of the optimization problem, leading to more efficient searches and robust solutions.

Another important feature of ES is its problem-dependent representation. Unlike other EA subclasses, ES uses natural representations, meaning the problem space and the search space are identical. This direct approach can lead to more intuitive solutions and simplifies the process of decoding the candidate solutions back into meaningful real-world answers.

The mathematical foundation of ES is also well-established, with concepts like the Covariance Matrix Adaptation (CMA) playing a significant role in guiding the evolution process. CMA helps in estimating the covariance of successful mutations, thereby aligning the search distribution with the shape of the problem landscape.

Historically, ES has its roots in the work of early computer scientists who applied evolutionary processes to engineering optimization problems. This led to the development of various strategies, each with its unique approach to variation and selection. Over the years, ES has been refined and expanded, incorporating advanced techniques such as derandomization and co-evolution.

In practice, ES has been applied to a wide range of fields, from engineering design to machine learning. Its ability to handle complex, high-dimensional, and non-linear problems makes it a versatile tool for researchers and practitioners alike.

As the field of evolutionary computation continues to grow, ES remains a vibrant area of research. Ongoing developments are focused on improving the efficiency and effectiveness of these strategies, ensuring that they remain at the forefront of optimization technology. The exploration of hybrid approaches, combining ES with other optimization methods, is also an exciting avenue, promising even more powerful and adaptable solutions for the challenges of tomorrow.

2.3 Particle swarm optimization

Particle Swarm Optimization (PSO) is a popular algorithm that falls under the category of nature-inspired heuristic optimization methods. It models the behavior of a swarm (such as birds or fish) collectively searching for an optimal solution by sharing information within the group [20]. Although commonly used, it is important to consider its convergence conditions, how it handles local optima, and the effects of its parameter settings on performance for successful implementation.

PSO Pseudocode:

Initialize the swarm, assigning random positions and velocities to each particle.
For each particle in the swarm:
  Set the particle's personal best position to its current position.
  If the particle's current position is better than its personal best:
    Update the particle's personal best.

Set the global best position as the best position among all particles in the swarm.

Repeat while the termination condition is not met:
  For every particle in the swarm:
    Adjust the velocity as follows:
      *velocity = inertia * velocity.*
        *+ cognitive_coefficient * random() * (personal_best_position - current_position).*
        *+ social_coefficient * random() * (global_best_position - current_position).*

  Update the particle's position:
    position = position + velocity.

  If the current position is better than the particle's personal best:
    Update the personal best.

  If the current position is better than the global best position:
    Update the global best.
Return the global best position as the solution.

PSO has three main considerations:

1. Convergence conditions: Local Optima Challenge, and Parameter Settings. Convergence Conditions, The algorithm stops when reaching a maximum number of iterations, minimal improvement in the global best position, or no improvement over a defined period to prevent endless execution.

2. Local optima challenge: To avoid local optima, strategies like inertia weighting, velocity clamping, and introducing randomness are used to enhance exploration.

3. Parameter settings: Key parameters—inertia weight, cognitive coefficient, social coefficient, and random values—impact performance. Proper tuning balances exploration and exploitation to prevent slow convergence or stagnation.

2.4 Ant colony optimization

Ant colony optimization (ACO), introduced by Marco Dorigo in 1992 [21] is a nature-inspired heuristic algorithm that mimics ants' foraging behavior to solve optimization problems. In ACO, artificial "ants" simulate pheromone-laying and following behaviors to efficiently discover optimal paths through graphs, much like real ants finding food. The ACO algorithm is effective for path-finding and graph-based problems, such as vehicle routing and scheduling, using a probabilistic approach to escape local optima. It is part of the swarm intelligence class, utilizing simple agents to create complex global behaviors, making it robust in dynamic environments. Key parameters of this algorithm include Pheromones, Evaporation rate, and Alpha and Beta. The algorithm converges when minimal Pheromone change appears or maximum iterations are reached, or the solution does not improve [22].

Pseudocode for Ant Colony Optimization:

Initialize pheromone levels on all paths.
Set parameters: α (pheromone influence), β (heuristic influence), ρ (evaporation rate), Q (pheromone deposit factor).
Repeat until stopping criteria met:
  For each ant in the colony:
    Construct a solution:
      Start from a random node.
      While solution is not complete:
        Select the next node probabilistically based on pheromone level (τ) and heuristic information (η).
        Add selected node to the solution.
  Evaluate the solution's quality.
Update pheromones on all paths:
  For each path:
    Evaporate pheromone: $\tau = (1 - \rho) * \tau$.
    Deposit new pheromone for paths used by ants based on solution quality.
If a new global best solution is found:
  Update the global best solution.
End
Return the best solution found.

2.5 Simulated annealing

Simulated Annealing (SA) is a probabilistic optimization technique inspired by the annealing process in metallurgy [23]. It is effective for complex combinatorial problems like the traveling salesman problem and job-shop scheduling due to its ability to escape local optima. SA begins with a high temperature to allow exploration, then gradually cools, reducing the acceptance of worse solutions and focusing on promising areas. This balance prevents premature convergence and enhances the likelihood of finding the global optimum. Like genetic algorithms and ant colony optimization, SA is a versatile metaheuristic, but the "No Free Lunch" theorem suggests no single algorithm is best for all problems.

Pseudocode for Simulated Annealing:

Initialize the current solution (current_solution).
Set initial temperature (T) and cooling rate (α).
Set the maximum number of iterations (max_iterations).

Set the stopping criterion (minimum temperature or max iterations).
Repeat until stopping criteria is met:
For each iteration:
Generate a neighboring solution (new_solution) from the current solution.
Calculate the change in objective value: ΔE = objective(new_solution) - objective(current_solution).

If new_solution is better (ΔE < 0):
Accept new_solution as the current solution.
Else:
Accept new_solution with a probability P = exp(-ΔE /T).

*Update temperature: T = α * T//Reduce the temperature.*
If current_solution is better than the best_solution found:
Update the best_solution.

Return the best_solution.

2.6 Artificial bee colony algorithm

The Artificial Bee Colony (ABC) Algorithm, introduced by Karaboga in 2005 [24], is a nature-inspired optimization technique simulating honeybee foraging. Known for its simplicity and effectiveness, it is similar to particle swarm optimization. Enhancements like the ABCDC variant improve convergence by dynamically adjusting the employed and onlooker bee ratios, while the CNSABC algorithm uses Bernoulli chaotic mapping and neighborhood search to boost diversity and search efficiency. These advancements show the ongoing refinement of metaheuristics for tackling complex optimization problems. ABC has been widely applied in engineering and finance, demonstrating the potential of bio-inspired approaches. Continued research aims to further evolve the algorithm, solidifying its role in modern optimization challenges.

Initialize the population of food sources (solutions).
Evaluate the fitness of each food source.
Set the initial parameters: number of employed bees, onlooker bees, and maximum iterations.
Repeat until the stopping criteria are met:
*// **Employed Bee Phase**.*
For each employed bee:
Select a food source (solution).
Generate a new solution by modifying the current one.
Calculate the fitness of the new solution.
If the new solution is better, replace the old solution.

*// **Onlooker Bee Phase**.*
Calculate the probability for each food source based on fitness.
For each onlooker bee:
Select a food source probabilistically using the fitness probability.
Generate a new solution and evaluate it.
If the new solution is better, replace the old solution.
*// **Scout Bee Phase**.*
If a food source has not improved for a predefined number of cycles:
Replace it with a randomly generated new solution (Scout Bee behavior).

Memorize the best solution found so far.
Update iteration counter.

Return the best solution found.

2.7 Quantum red fox optimization (QRFO)

Quantum Red Fox Optimization is a quantum-based algorithm which integrates the hunting behavior of red foxes with principles of quantum computing principles like entanglement and superposition for improved search results and efficiency. This algorithm terminates when the solution has stabilized over several iterations, when the variance of fitness values across the population is below a defined threshold, or when a predefined number of iterations is reached. The pseudocode of this algorithm can be found below:

1: Initialize population of foxes with random positions and momenta.
2: Define the objective function f (x).
3: while termination condition not met do.
4: for each fox in the population do.
5: Update position and momentum using quantum operators.
6: Calculate the fitness f x) of the fox.
7: Apply quantum superposition to explore multiple positions.
8: Update the best-known position if current position is better.
9: end for.
10: Apply quantum interference to enhance exploitation.
11: Update global best position.
12: end while.
13: Output the best-found solution

The use of quantum superposition makes this algorithm less prone to being trapped in local optima because it allows the algorithm to explore multiple solutions at the same time. Parameters of the algorithm include quantum bit state representation, hunting intensity (which defines how aggressive the search is), and fox agents' step size. Managing and tuning these parameters is the key to the effectiveness of this algorithm and avoiding slow convergence or ineffective search [25].

2.8 Random initialization with memory enhancement (RIME)

Rime uses memory-based mechanisms that allow the algorithms to recall successful strategies from previous optimization while maintaining diversity through random initialization. The algorithm converges when the fitness function does not improve or a predetermined number of iterations is reached. To escape getting trapped in local optima, RIME uses the memory mechanism to revisit the parts of the search space with good potential. However, it may still get stuck in suboptimal solutions if the memory threshold is too low. The main parameters which affect the algorithm are memory retention factor and frequency of memory updates. Incorrect settings of these parameters can lead to over-exploration or too much reliance on past solutions. The pseudocode of this algorithm is as follows [26]:

1: Initialize the rime population R.
2: Get the current optimal agent and optimal fitness.
3: While $t < T$.
4: Coefficient of adherence $E = (t/T)0.5$.

5: *if r2 < E.*
Update rime agent location by the soft-rime search strategy.
End if.
6: *if r3 < NormalizefitnessofSi.*
Cross updating between agents by the hard-rime puncture mechanism.
End if.
7: *if F(Rinew) < F(Ri).*
Select the optimal solution and replace the suboptimal solution.
using the positive greedy selection mechanism.
End if.
8: *t = t + 1.*
9: *End while*

Having explored the fundamental concepts of evolutionary algorithms and heuristic optimization techniques, we now turn our attention to their practical applications in automotive engineering. These powerful tools have revolutionized various aspects of vehicle design and manufacturing, offering innovative solutions to complex problems that traditional methods struggle to address.

3. Integration of evolutionary algorithms in automotive design

3.1 Vehicle design optimization

3.1.1 Aerodynamics optimization

Vehicle aerodynamics optimization aims to reduce air resistance, enhancing fuel efficiency and performance. Computational Fluid Dynamics (CFD) advancements have improved airflow simulation, leading to effective designs [27]. Aerodynamic accessories like air dams and spoilers reduce drag, improving stability. For electric vehicles (EVs), optimizing aerodynamics is crucial, as over 50% of energy at highway speeds is used to overcome air resistance, impacting battery weight, cost, and range. Innovations like honeycomb structures and strategic fins have been shown to improve EV range. Methods like RKri-EGO-PEI filter noise from CFD simulations ensure smoother optimization. Hybrid techniques, combining gradient-based and gradient-free methods, enhance understanding of aerodynamic forces, enabling designs with lower drag coefficients and better performance.

3.1.2 Lightweight design and material selection

Integrating evolutionary algorithms in automotive design, especially for lightweight material selection, addresses the industry's focus on efficiency and sustainability [28]. Recent research optimizes components like car door panels using materials such as steel alloy, aluminum alloy, magnesium alloy, and Carbon Fiber Reinforced Composite (CFRP) due to their high strength-to-weight ratios, crucial for reducing fuel consumption and CO_2 emissions [29, 30]. Genetic algorithms like GANetXL effectively identify optimal materials while preserving mechanical properties. A new performance-driven lightweight optimization method combines multi-material selection and multi-objective optimization, handling material discreteness and nonlinearity. It incorporates constraints like static stiffness and crashworthiness, achieving mass and cost reductions in parts such as the cab's skeleton beam.

Additionally, hybrid approaches integrating machine learning and evolutionary strategies are emerging for structural optimization and material selection, navigating complex material properties and design requirements. An eco-design tool using genetic algorithms optimizes the co-selection of materials and joining methods, considering environmental impacts. This tool represents progress in sustainable design by aligning material and manufacturing choices with ecological goals. These advancements highlight the potential of combining computational models and evolutionary algorithms to create efficient, lightweight, and eco-friendly automotive designs.

3.1.3 Engine performance tuning

Engine performance tuning leverages engineering expertise and advanced mathematical optimization techniques to enhance efficiency and output, minimizing the need for extensive physical testing [31]. Methods like the Wiebe function and Miyamoto model help assess combustion rates and engine characteristics, providing fast and cost-effective tuning [32]. Optimization techniques vary widely:

1. *Response surface methodology (RSM)* [33]: A statistical approach that models the effects of various tuning parameters on engine performance.

2. *Genetic algorithms (GAs)*: Mimic natural evolution to solve complex optimization problems, both constrained and unconstrained.

3. *Particle swarm optimization (PSO)*: Uses a population of candidate solutions (particles) that move through the search space to optimize engine parameters.

4. *Proportional-integral-derivative (PID) controllers*: Applied in fuel system optimization, PID controllers use feedback loops to reach setpoints. Tuning methods include Ziegler-Nichols, Cohen-Coon, and auto-tuning software.

5. *Multi-objective optimization (MOO)*: Useful for balancing conflicting goals like fuel efficiency and power. Techniques like Pareto optimization find a set of optimal solutions (Pareto front).

These techniques, when used effectively, lead to enhanced fuel efficiency, increased power output, and better engine responsiveness. As engine tuning evolves, these mathematical approaches will remain crucial for developing high-performance engines, showcasing the value of combining engineering knowledge with computational models and optimization strategies.

The application of EAs and heuristic optimization in vehicle design has led to significant improvements in performance, efficiency, and sustainability. By optimizing aerodynamics, materials, and engine performance, engineers can create vehicles that meet increasingly stringent environmental regulations while enhancing driver experience [34].

3.2 Optimization in automotive manufacturing processes

3.2.1 Assembly line balancing

Assembly line balancing is a critical aspect of automotive manufacturing processes, where the goal is to distribute tasks among workstations in a way that maximizes efficiency and minimizes production time. Recent studies have focused on developing multi-objective models that consider not only the total time and line efficiency but also customer satisfaction indices. For instance, a model known as SLBP-2E combines Simple Line Balancing Problems Type 2 and Type E, addressing them with heuristic solution approaches due to their non-polynomial hard formulation [35]. Moreover, the classification and balancing of assembly lines have been explored to optimize the segmentation of tasks with objectives like minimizing cycle time and maximizing profit. Simulation techniques have also been employed to improve assembly line performance, which can be more cost-effective than traditional trial-and-error methods. Additionally, research has been conducted on the balance of assembly lines by proposing dual-objective optimization models and using simulated annealing to solve complex balancing problems. These advancements in assembly line optimization underscore the importance of robust and adaptive manufacturing systems that can respond to the dynamic demands of the market. The integration of simulation-based layout optimization and heuristic approaches signifies a progressive step toward enhancing the agility and efficiency of automotive manufacturing processes.

3.2.2 Supply chain optimization

Automotive supply chains are moving from lean strategies to more resilient models, emphasizing risk mitigation through dual sourcing, strategic warehousing, and "right-shoring" to optimize cost, efficiency, and stability [36].

Automotive companies are prioritizing supply chain resilience due to COVID-19 and semiconductor shortages by improving monitoring, predictive modeling, and crisis management. Effective management now requires integrating demand forecasting, lean manufacturing, and inventory optimization. Non-linear optimization models are essential for aligning production and inventory at strategic and operational levels. The shift toward electric vehicles further complicates supply chains, requiring new risk management strategies and supplier integration [37].

Addressing supply chain bottlenecks requires solutions like collaboration, supplier diversification, and material sourcing. These strategies help the automotive industry enhance efficiency, adaptability, and resilience in a complex supply chain ecosystem.

3.2.3 Quality control and defect minimization

Optimization in automotive quality control focuses on enhancing vehicle reliability and safety while reducing costs. Recent advancements include using manufacturability-constrained optimization with recycled carbon fiber-metal hybrids for lightweight structures, improving part quality, and reducing cycle time and warpage. Hybrid optimization methods, such as combining neural networks with genetic algorithms, optimize plastic injection molding, minimizing waste and energy consumption. Proactive quality control strategies and continuous improvements are essential to addressing common defects and deviations in automotive manufacturing [38]. Automotive manufacturers are using new technologies and lean principles to produce safer, more reliable vehicles efficiently. Combining advanced materials, smart designs, and optimization techniques is driving the production of next-generation vehicles that meet high quality and environmental standards.

The integration of EAs and heuristic optimization into manufacturing processes has transformed the automotive industry's approach to production. From streamlining assembly lines to optimizing supply chains and enhancing quality control, these techniques have enabled manufacturers to increase efficiency, reduce costs, and improve product quality [39].

3.3 Advanced automotive technologies

3.3.1 Electric vehicle battery design

Optimization techniques in electric vehicle (EV) battery design are crucial for enhancing performance, reducing costs, and improving sustainability [40–42]. A specific example of such an application is the use of generative design and origami-inspired topological design enabled by additive manufacturing. These advanced techniques allow for the creation of battery components with optimized thermal, electrical, and mechanical properties. Additionally, sensitivity studies of battery performance with alternate materials can be conducted to identify the best combinations for efficiency and longevity. Another example is the optimization strategy based on the convex optimization algorithm, which has been shown to improve the motor model and power battery parameters of EVs or hybrid electric vehicles [43]. This approach not only enhances the vehicle's performance but also contributes to lower energy consumption and reduced environmental impact. Furthermore, multiphysics simulation optimization frameworks are employed to design lithium-ion battery packs, ensuring that they meet the specific requirements for EV applications. Lastly, the design optimization of the battery pack enclosure is addressed through methodologies that minimize mass and deformation while maximizing natural frequency [44]. These examples highlight the interdisciplinary nature of EV battery design optimization, encompassing power management, system integration, and vehicle dynamics, among others. The ongoing research and development in this field are vital for the advancement of EV technology and its adoption worldwide.

3.3.2 Autonomous vehicle navigation systems

Optimizing autonomous vehicle navigation is a complex challenge focused on enhancing path-finding efficiency, safety, and reliability [45]. Recent advancements include a hybrid path planning approach combining adaptive bidirectional A*, vector triangulation, and path smoothing to improve trajectory smoothness and minimize sharp turns, enhancing vehicle stability and passenger comfort [46]. Integrated perception and planning frameworks using LiDAR and inertial sensor fusion with Error-State Extended Kalman Filters (ES-EKF) generate real-time cost maps for precise state estimation and navigation in cluttered environments. AI-driven algorithms like A* search, Dijkstra's algorithm, and genetic algorithms optimize routes by considering traffic, road rules, and obstacles [47].

Reinforcement learning (RL) using Generalized Advantage Estimation (GAE) is also applied to navigate complex environments, utilizing the actor-critic framework for continuous learning [48]. SLAM-based frameworks integrate ES-EKF for sensor fusion, solving real-time trajectory optimization problems. Vision-aided systems employ deep learning for depth and object-level perception, enabling efficient navigation in dynamic scenarios. These techniques improve both individual vehicle performance and broader traffic management, contributing to safer and more efficient urban mobility. As autonomous vehicles become more prevalent, refining these optimization strategies will shape the future of transportation and smart cities.

3.3.3 Safety feature enhancements

Heuristic optimization methods, which rely on experience, intuition, and trial-and-error, are valuable for enhancing safety features in complex systems where traditional techniques may not suffice. These methods prioritize adaptability and speed over finding the perfect solution, making them ideal for developing robust and resilient safety systems. In transportation, heuristics improve autonomous vehicle safety by enabling quick decision-making, while in industrial settings, they optimize factory layouts to reduce accident risks and enhance worker safety. This adaptability makes heuristics a practical choice for safety-related optimization challenges [49].

Multi-objective heuristic optimization addresses conflicting objectives like cost versus safety, offering balanced solutions for real-world safety projects, such as level crossing improvements. Advanced methods, including quantum-inspired and hyper-heuristic approaches, further enhance safety features by dynamically adapting parameters and exploring broader solution spaces. These sophisticated techniques contribute to more effective safety optimizations [49].

The ongoing evolution of nature-inspired heuristic methods, supported by cutting-edge research, ensures continuous advancements in safety standards across industries. As these techniques are refined and more widely adopted, they will significantly improve safety measures, enhancing protection for individuals and communities worldwide. This progress underscores the value of modern heuristic optimization in achieving robust safety outcomes [49].

Heuristic approaches in safety optimization address complex problems where traditional methods fall short. Pareto Multi-objective Optimization [50] is used to balance conflicting objectives like cost and safety, providing a set of optimal solutions for informed decision-making. Algorithms for safety improvements at level crossings consider costs, effectiveness, and externalities (e.g., traffic delays) to prioritize safety measures. Other techniques include Elimination-by-Aspects, which discards options based on disqualifying traits, and the Recognition heuristic, using past experiences to inform decisions. The Availability heuristic evaluates risk based on the ease of recalling similar events, making these methods practical tools for safety-related decision-making and risk assessment.

Hyper-heuristic approaches, often using reinforcement learning, dynamically adapt to problem domains by rewarding improvements, making them effective for enhancing safety in advanced systems like aircraft structural designs. The continuous evolution of heuristic methods, inspired by nature and technology, drives advancements in safety optimization, contributing to more robust and resilient safety features across industries. Integrating these approaches into safety systems showcases progress in creating reliable mechanisms that safeguard individuals and communities. For more insights, resources on heuristic optimization applications, including power systems and hybrid methods, provide an in-depth understanding of modern techniques and their practical uses [51].

As the automotive industry moves toward electrification and autonomy, EAs and heuristic optimization play a crucial role in developing advanced technologies. These techniques are instrumental in designing efficient batteries, creating robust navigation systems for autonomous vehicles, and enhancing safety features to protect passengers and pedestrians alike.

To further illustrate the impact of evolutionary algorithms and heuristic optimization in automotive engineering, let us examine some real-world case studies that demonstrate the practical application of these techniques.

4. Case studies

4.1 Application of improved particle swarm optimization for navigation of unmanned surface vehicles

The paper "Application of Improved Particle Swarm Optimization for Navigation of Unmanned Surface Vehicles" by Junfeng Xin et al. introduces an enhanced PSO algorithm to address limitations like premature convergence and fixed parameters in USV navigation. The authors propose three strategies: linearly descending inertia weight, adaptively controlled acceleration coefficients, and random grouping inversion, which improve path planning by enhancing robustness, minimizing path lengths, and preventing premature convergence. Monte Carlo simulations and real-world tests show that the refined PSO yields shorter and more efficient paths. Adaptively controlled acceleration significantly reduces path length, while linearly descending inertia enhances robustness. Random grouping maintains diversity and accelerates convergence. The findings demonstrate superior performance over traditional methods, offering valuable insights for future autonomous marine navigation applications (**Figure 1**) [52].

4.2 Speed control of hybrid electric vehicles using genetic algorithm

Prerna Gaur et al. applied a Genetic Algorithm (GA) to optimize speed control in Hybrid Electric Vehicles (HEVs) by fine-tuning a Proportional-Integral-Derivative (PID) controller. Their GA-PID controller outperformed conventional methods like Ziegler-Nichols and manual tuning, showing reduced overshoot, faster settling time, and improved stability, making it an effective solution for enhancing HEV efficiency and safety [53].

4.3 Optimizing the active suspension system of automobiles using a reduced-order approach

Sairoel Amertet Finecomes and his team optimized active suspension systems in vehicles using a linear quadratic regulator (LQR) controller with actuator dynamics, applied to a reduced-order car model. Their results showed that incorporating

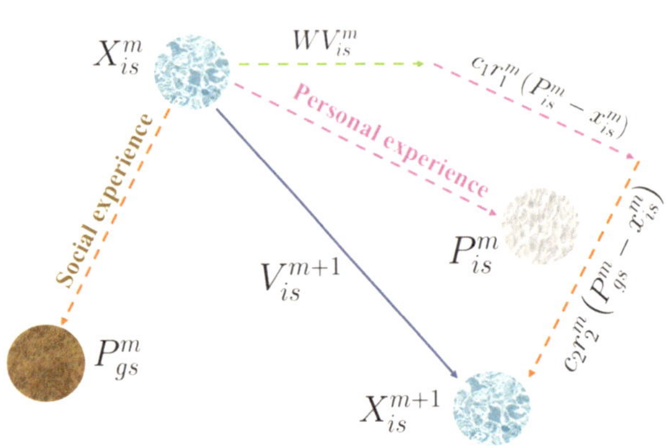

Figure 1.
Schematic diagram of a particle's position update in PSO [52].

actuator dynamics improved vehicle stability, reduced vertical displacement by 25%, and enhanced velocity control by 38.46% compared to a conventional passive system, making it a more effective solution for suspension performance [54].

4.4 Improved genetic algorithm for the granularity-based Split vehicle routing problem with simultaneous delivery and pickup

Yuxin Liu, Zihang Qin et al. conducted a study focusing on solving the Granularity-based Split Vehicle Routing Problem with Simultaneous Delivery and Pickup (GSVRPSDP). The researchers developed a Genetic-Simulated hybrid algorithm (GA-SA) by integrating the global search ability of Genetic Algorithms (GA) with the local optimization capacity of Simulated Annealing (SA). Their proposed algorithm significantly reduced transportation costs, achieving over 10% lower costs compared to conventional methods such as GA, SA, and Particle Swarm Optimization (PSO). Moreover, the GA-SA approach demonstrated higher space and capacity utilization, making it a robust solution for complex vehicle routing problems [55].

4.5 Optimization design of automotive body stiffness using a boundary hybrid genetic algorithm

This case study applies a Boundary Hybrid Genetic Algorithm (BHGA) to optimize the stiffness of a Body-in-White (BIW) automotive structure during the conceptual design stage. Using the Thin-Walled Beam Transfer Matrix Method (TBTMM) for accurate mechanical modeling, BHGA integrates global and local search to achieve weight reduction without compromising stiffness. BHGA outperformed other metaheuristic algorithms, avoiding local optima and meeting engineering constraints, making it ideal for early design stages. This approach provides a practical solution for achieving lightweight, safe BIW structures, showcasing its effectiveness in real-world automotive design applications (**Figure 2**) [56].

5. Challenges and future directions

As we look to the future of automotive engineering, several challenges and opportunities emerge in the application of evolutionary algorithms and heuristic optimization techniques.

5.1 Scalability and complexity issues

In the realm of automotive engineering, the application of evolutionary algorithms and nature-inspired heuristic optimization has become increasingly vital due to the complex, multi-dimensional nature of design and manufacturing processes. These computational strategies, inspired by biological evolution and natural phenomena, offer robust solutions to optimization problems that are otherwise too complex for traditional methods. However, scalability and complexity remain significant challenges. As automotive systems become more sophisticated, the need for algorithms that can efficiently navigate vast solution spaces and handle numerous variables grows. The integration of these algorithms into automotive engineering requires careful consideration of computational resources, convergence rates, and the ability to find global optima in highly complex landscapes. Researchers are actively exploring ways

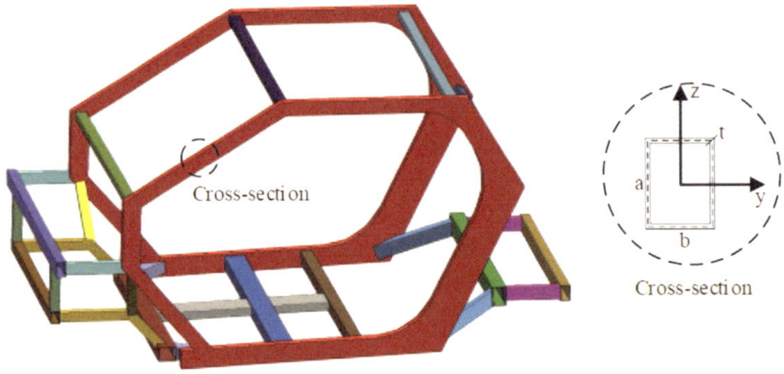

Figure 2.
The conceptual BIW model [56].

to enhance the scalability of these algorithms, such as through parallel computing and adaptive mechanisms that adjust parameters in real-time. Moreover, the complexity of automotive engineering problems often necessitates a hybrid approach, combining the strengths of various nature-inspired algorithms to achieve the best results. As these methods continue to evolve, they hold the promise of driving innovation in automotive design, leading to vehicles that are not only more efficient and reliable but also better adapted to the changing demands of technology and society [57].

5.2 Integration with machine learning and AI

The integration of Machine Learning (ML) and Artificial Intelligence (AI) into evolutionary algorithms and nature-inspired heuristic optimization is revolutionizing automotive engineering [58]. These advanced computational techniques are enhancing the development of autonomous vehicles, optimizing design processes, and improving safety features. For instance, Optimization algorithms, particularly those driven by artificial intelligence (AI), have been applied to optimize V2G operations. These algorithms address the dynamic nature of energy demand, grid constraints, and user preferences. Case studies demonstrate the efficacy of AI-driven algorithms in improving grid stability, cost optimization, and user satisfaction [59, 60]. Evolutionary algorithms, inspired by natural selection, are particularly effective in optimizing complex engineering problems where multiple objectives must be balanced, such as fuel efficiency, cost, and performance. Nature-inspired heuristics, which mimic biological processes like swarm intelligence, are also being employed to find optimal solutions in large, dynamic search spaces typical in automotive design. These methods are not only accelerating the pace of innovation but also ensuring that vehicles meet the highest standards of safety and reliability. As the industry moves toward fully autonomous vehicles, the role of AI and ML in evolutionary algorithms will become increasingly central, driving forward the future of transportation.

5.3 Ethical considerations in automated design

The integration of evolutionary algorithms and nature-inspired heuristic optimization in automotive engineering presents many opportunities, but also raises important ethical concerns. As these technologies increasingly influence vehicle design, ensuring their responsible use is essential. This section highlights the key ethical considerations: safety, transparency, bias, environmental impact, job displacement, and accountability.

5.3.1 Safety and reliability

Safety is paramount in automotive design, especially with automated processes. Evolutionary algorithms optimize vehicle designs for performance, but this can sometimes inadvertently compromise safety. For example, systems designed to enhance aerodynamics or reduce material usage could weaken structural integrity or reduce the effectiveness of safety features [61].

Automated design systems must be equipped with constraints to ensure that safety-critical elements, like braking or steering, are not compromised. Moreover, relying solely on simulations can miss edge cases that occur in real-world scenarios. Therefore, thorough testing beyond digital environments is essential to ensure vehicle safety and reliability [61].

5.3.2 Transparency

A major concern with advanced algorithms is their "black box" nature, where decision-making processes are not easily understood. This lack of transparency is problematic, especially when safety-critical or regulatory decisions are based on algorithmic outputs [61].

For public trust and regulatory compliance, it is crucial that manufacturers can explain how design decisions are made by these systems. Ethical use of automated design requires developing algorithms that are interpretable and ensure that human engineers understand the rationale behind key design choices [61].

5.3.3 Bias and fairness

Algorithms can reflect biases based on the data they are trained on or the design goals set by developers. If an algorithm optimizes for cost, for example, it might result in designs that exclude certain groups, such as people with disabilities or those requiring accessible features [61].

To address this, diverse datasets should be used, and inclusivity should be an explicit design goal. Additionally, teams working with these systems need to represent a broad range of perspectives to avoid creating designs that unintentionally marginalize certain users [61].

5.3.4 Environmental impact

Automated design can optimize vehicles for fuel efficiency or material reduction, but it may also favor solutions that are harmful to the environment, such as the use of non-recyclable materials or rare earth metals. Ethical design requires that sustainability be built into the optimization process [62].

Environmental impact should be a core consideration, not only in vehicle performance but throughout the entire lifecycle, from material sourcing to production and disposal. Designers must ensure that the algorithms they use account for sustainability metrics alongside performance [62].

5.3.5 Job displacement

As automated design processes become more widespread, they may displace human designers, engineers, and production workers. While automation can increase efficiency, it risks replacing roles traditionally reliant on human expertise [63].

Ethically, the automotive industry must balance technological advancements with the well-being of its workforce. This includes reskilling workers to collaborate with automated systems and preparing the workforce for shifts in required skills, ensuring they are not left behind by these innovations [63].

5.3.6 Accountability and liability

When vehicles designed by algorithms malfunction, assigning responsibility becomes complicated. Was it a flaw in the algorithm, a misjudgment by the engineers, or a manufacturing issue? Diffusion of responsibility can blur accountability, which is ethically problematic, particularly in the event of accidents [64].

Clear accountability measures must be established throughout the automated design process, ensuring that human oversight is maintained, and that algorithms are transparent enough to allow for tracing decisions back to their source. Regulatory frameworks must also evolve to address these new challenges [64].

6. Conclusion

Evolutionary algorithms (EAs) and heuristic optimization techniques have significantly impacted automotive engineering by offering robust solutions to complex design and manufacturing challenges. Inspired by natural selection, these methods provide adaptive approaches for optimizing vehicle performance, fuel efficiency, and structural integrity. Automakers like Renault SA have utilized them to reduce vehicle weight through innovative material selection while maintaining safety standards. This flexibility extends to advancements in autonomous vehicle technology and energy-efficient designs, enabling novel solutions beyond deterministic methods. As the automotive industry moves toward electrification and autonomy, EAs will play a central role in addressing sustainability and safety challenges.

However, EAs and heuristics face limitations, such as high computational demands, slow response times, and sensitivity to parameter settings. Real-time optimization and multi-objective balancing (e.g., performance vs. efficiency) remain challenging, especially for high-dimensional problems. They also struggle in dynamic environments and large-scale applications due to scalability issues. To overcome these constraints, hybrid approaches and algorithmic modifications are essential to improve adaptability and efficiency, ensuring these techniques can meet the evolving demands of the automotive industry. As AI and large-scale models are integrated, EAs will unlock new possibilities for personalized, efficient, and safer vehicles, shaping the future of transportation.

Author details

Morteza Mollajafari[1]* and Salman Ebrahimi-Nejad[2]

1 Vehicle Electrical and Electronic Research Lab, School of Automotive Engineering, Iran University of Science and Technology, Tehran, Iran

2 Vehicle Dynamical Systems Research Lab, School of Automotive Engineering, Iran University of Science and Technology, Tehran, Iran

*Address all correspondence to: mollajafari@iust.ac.ir

IntechOpen

© 2025 The Author(s). Licensee IntechOpen. This chapter is distributed under the terms of the Creative Commons Attribution License (http://creativecommons.org/licenses/by/4.0), which permits unrestricted use, distribution, and reproduction in any medium, provided the original work is properly cited.

References

[1] Stańczak JT. Efficient selection methods in evolutionary algorithms. Computer Science. 2024;**25**(1). DOI: 10.7494/csci.2024.25.1.5330

[2] Wang Z, Pei Y, Li J. A survey on search strategy of evolutionary multi-objective optimization algorithms. Applied Sciences. 2023;**13**(7):4643

[3] de Sousa MC, Meneghini IR, Guimarães FG. Assessment of robust multi-objective evolutionary algorithms on robust and noisy environments. In: Naldi MC, Bianchi RAC, editors. Intelligent Systems. BRACIS 2023, Lecture Notes in Computer Science. Vol. 14197. Cham: Springer; 2023. DOI: 10.1007/978-3-031-45392-2_3

[4] IHMSC 2022 Committees. 2022 14th International Conference on Intelligent Human-Machine Systems and Cybernetics (IHMSC). Hangzhou, China: IEEE; 2022. pp. 12-13. DOI: 10.1109/IHMSC55436.2022.00006

[5] Langdon WB. Genetic programming convergence. Genetic Programming and Evolvable Machines. 2022;**23**:71-104. DOI: 10.1007/s10710-021-09405-9

[6] Yan WQ, Yang ZY, Yu YY, Lin Y. A hybrid algorithm and collaborative optimization strategy based on novel coding method for SPRD. Ocean Engineering. 2024;**311**:118911

[7] Layeb A. Differential evolution algorithms with novel mutations, adaptive parameters, and Weibull flight operator. Soft Computing. 2024;**28**:7039-7091. DOI: 10.1007/s00500-023-09561-3

[8] Tan J, Chen H, Ye X, Zhai G, Yan J. A heuristic robust design optimization approach for batch production permanent magnet actuator consider various uncertainties and life-cycle quality loss. IEEE Transactions on Energy Conversion. 2024;**39**(2):805-816. DOI: 10.1109/TEC.2023.3332915

[9] Zhao X, Li J, Sun S, Han C, Zhu W, He Z, et al. Toward structure optimization for the mobile vehicle system based on multiconstraints. Robotic Intelligence and Automation. 2023;**43**(1):75-84. DOI: 10.1108/RIA-08-2022-0213

[10] Oroun Gobi I, Guo CG. Finite element analysis and automation of a medium scale grinder applied to the manufacture of cassava starch. Frontiers of Mechanical Engineering. 2024;**10**:1427388. DOI: 10.3389/fmech.2024.1427388

[11] Karimi-Mamaghan M, Mohammadi M, Meyer P, Karimi-Mamaghan AM, Talbi EG. Machine learning at the service of meta-heuristics for solving combinatorial optimization problems: A state-of-the-art. European Journal of Operational Research. 2022;**296**(2):393-422

[12] Lu J, Hu W, Wang Y, Li L, Ke P, Zhang K. A hybrid algorithm based on particle swarm optimization and ant colony optimization algorithm. In: Qiu M, editor. Smart Computing and Communication. SmartCom 2016, Lecture Notes in Computer Science. Vol. 10135. Cham: Springer; 2017

[13] Chen S, Hu X, Zhao J, Wang R, Qiao M. A review of decision-making and planning for autonomous vehicles in intersection environments. World Electric Vehicle Journal. 2024;**15**:99. DOI: 10.3390/wevj15030099

[14] Keoleian GA, Sullivan JL. Materials challenges and opportunities for

enhancing the sustainability of automobiles. MRS Bulletin. 2012;**37**:365-373. DOI: 10.1557/mrs.2012.52

[15] Li N, Ma L, Yu G, Xue B, Zhang M, Jin Y. Survey on evolutionary deep learning: Principles, algorithms, applications and open issues. ACM Computing Surveys. 2022;**56**:1-34

[16] Eiben AE, Smith JE. Introduction to Evolutionary Computing. 2nd ed. Berlin, Heidelberg: Springer Verlag; 2015. DOI: 10.1007/978-3-662-44874-8

[17] Alhijawi B, Awajan A. Genetic algorithms: Theory, genetic operators, solutions, and applications. Evolutionary Intelligence. 2024;**17**:1245-1256. DOI: 10.1007/s12065-023-00822-6

[18] Gorji SA. Challenges and opportunities in green hydrogen supply chain through metaheuristic optimization. Journal of Computational Design and Engineering. 2023;**10**(3):1143-1157. DOI: 10.1093/jcde/qwad043

[19] Emmerich M, Shir OM, Wang H. Evolution strategies. In: Martí R, Panos P, Resende M, editors. Handbook of Heuristics. Cham: Springer; 2018. DOI: 10.1007/978-3-319-07153-4_13-1

[20] Gad AG. Particle swarm optimization algorithm and its applications: A systematic review. Archives of Computational Methods in Engineering. 2022;**29**(5):2531-2561

[21] Dorigo M. Optimization, learning and natural algorithms [thesis]. Italy: Politecnico di Milano; 1992

[22] Dorigo M, Stützle T. Ant colony optimization. Scholarpedia. 2004;**2**(3):1461

[23] Delahaye D, Chaimatanan S, Mongeau M. Simulated annealing: From basics to applications. In: Gendreau M, Potvin JY, editors. Handbook of Metaheuristics. International Series in Operations Research & Management Science. Vol. 272. Cham: Springer; 2019. DOI: 10.1007/978-3-319-91086-4_1

[24] Zhang J, Wang H, Wang X. Application of artificial bee colony algorithm based on homogenization mapping and collaborative acquisition control in network communication security. PLoS One. 2024;**19**(7):e0306699. DOI: 10.1371/journal.pone.0306699

[25] Maskeliūnas R, Damaševičius R. Quantum red fox optimisation algorithm for supply chain forecasting in timber industry. In: 2024 6th International Conference on Computing and Informatics (ICCI). New Cairo - Cairo, Egypt: IEEE; 2024. pp. 361-365. DOI: 10.1109/ICCI61671.2024.10485104

[26] Hang S, Zhao D, Heidari AA, Liu L, Zhang X, Mafarja M, et al. RIME: A physics-based optimization. Neurocomputing. 2023;**532**:183-214. ISSN 0925-2312. DOI: 10.1016/j.neucom.2023.02.010

[27] Nath DS, Pujari PC, Jain A, et al. Drag reduction by application of aerodynamic devices in a race car. Advances in Aerodynamics. 2021;**3**:4. DOI: 10.1186/s42774-020-00054-7

[28] Xu W, Han H, Li Q, Mollajafari M, Scott F. Layerwise formulation of poroelastic composite plate under pre-buckling and thermal shock loading. Composite Structures. 2023;**304**:116343

[29] Zhang H, Sun Z, Zhi P, Wang W, Wang Z. Material-structure integrated design and optimization of a carbon-fiber-reinforced composite car door. Applied Sciences. 2024;**14**(2):930

[30] Ravanbod M, Ebrahimi-Nejad S, Mollajafari M. A thin-walled cavity structure with double-layer tapered scatterer locally resonant metamaterial plates for extreme low-frequency attenuation. International Journal of Solids and Structures. 2024;**293**:112742

[31] Ravanbod M, Ebrahimi-Nejad S, Mollajafari M, Salehi Rad B. Porous liner coated inlet duct: A novel approach to attenuate automotive turbocharger inlet flow-induced sound propagation. Engineering Research Express. 2023;**5**(1):015047

[32] Ghojel JI. Review of the development and applications of the Wiebe function: A tribute to the contribution of Ivan Wiebe to engine research. International Journal of Engine Research. 2010;**11**(4):297-312

[33] Korondi PZ, Marchi M, Poloni C. Response surface methodology. In: Vasile M, editor. Optimization under Uncertainty with Applications to Aerospace Engineering. Cham: Springer; 2021. DOI: 10.1007/978-3-030-60166-9_12

[34] Liu J, Zhang Q, Pei J, et al. fSDE: Efficient evolutionary optimisation for many-objective aero-engine calibration. Complex & Intelligent Systems. 2022;**8**:2731-2747. DOI: 10.1007/s40747-021-00374-1

[35] Khan AS, Khan R, Saleem W, Salah B, Alkhatib S. Modeling and optimization of assembly line balancing type 2 and E (SLBP-2E) for a reconfigurable manufacturing system. Processes. 2022;**10**(8):1582

[36] Rahman T, Paul SK, Agarwal R, Sarker R. Overview of supply chain risk and disruption management tools, techniques, and approaches. In: Paul SK, Agarwal R, Sarker RA, Rahman T, editors. Supply Chain Risk and Disruption Management. Flexible Systems Management. Singapore: Springer; 2023. DOI: 10.1007/978-981-99-2629-9_1

[37] Alicke K, Azcue X, Barriball E. Supply-Chain Recovery in Coronavirus Times—Plan for Now and the Future. McKinsey & Company; 2020. Available from: https://www.mckinsey.com/business-functions/operations/our-insights/supplychain-recovery-in-coronavirus-times-plan-for-now-and-the-future

[38] Maria-Diana E, Dumitrița-Ioana A, Marius VM. Quality control basic tools in automotive industry. In: Moldovan L, Gligor A, editors. The 16th International Conference Interdisciplinarity in Engineering. Inter-Eng 2022, Lecture Notes in Networks and Systems. Vol. 605. Cham: Springer; 2023. DOI: 10.1007/978-3-031-22375-4_44

[39] Beste D. Self-learning algorithms in automotive production. ATZ Electronics Worldwide. 2023;**18**:36-41. DOI: 10.1007/s38314-023-1452-x

[40] Shojaeefard MH, Mollajafari M, Pishe NE, Mousavi SM. Plug-in fuel cell vehicle performance and battery sizing optimization based on reduced fuel cell energy consumption and waste heat. Sustainable Energy Technologies and Assessments. 2023;**56**:103099

[41] Mesdaghi A, Mollajafari M. Improve performance and energy efficiency of plug-in fuel cell vehicles using connected cars with V2V communication. Energy Conversion and Management. 2024;**306**:118296

[42] Mollajafari M, Mohammadiazar F, Akbari E, Sheykhi E, Chaoui H. A new voltage-multiplier-based power converter configuration suitable

for renewable energy sources and sustainability applications. Sustainability. 2022;**14**(24):16698

[43] Mollajafari M, Kouhyar F. Simultaneous optimization of fuel and electrical energy consumption in forward-looking hybrid electric vehicle. Automotive Science and Engineering. 2022;**12**(1):3750-3761

[44] Shui L, Chen F, Garg A, et al. Design optimization of battery pack enclosure for electric vehicle. Structural and Multidisciplinary Optimization. 2018;**58**:331-347. DOI: 10.1007/s00158-018-1901-y

[45] Shojaeefard MH, Mollajafari M, Ebrahimi-Nejad S, Tayebi S. Weather-aware fuzzy adaptive cruise control: Dynamic reference signal design. Computers and Electrical Engineering. 2023;**110**:108903

[46] Shojaeefard MH, Mollajafari M, Mousavitabar SHR, Khordehbinan MW, Hosseinalibeiki H. A TSP-based nested clustering approach to solve multi-depot heterogeneous fleet routing problem. Métodos numéricos para cálculo y diseño en ingeniería: Revista internacional. 2022;**38**(1):1-11

[47] Bayliss C. Machine learning based simulation optimisation for urban routing problems. Applied Soft Computing. 2021;**105**:107269

[48] Mohtavipour SM, Ehsan TZ, Abeshoori HJ, et al. Smooth longitudinal driving strategy with adjustable nonlinear reference model for autonomous vehicles. International Journal of Dynamics and Control. 2023;**11**:2320-2334. DOI: 10.1007/s40435-023-01142-4

[49] Chang CCW, Ding TJ, Ee CCW, et al. Nature-inspired heuristic frameworks trends in solving multi-objective engineering optimization problems. Archives of Computational Methods in Engineering. 2024;**31**:3551-3584. DOI: 10.1007/s11831-024-10090-x

[50] Mohammadi A, Custódio AL. A trust-region approach for computing Pareto fronts in multiobjective optimization. Computational Optimization and Applications. 2024;**87**:149-179. DOI: 10.1007/s10589-023-00510-2

[51] Allen JG, Coates G, Trevelyan J. A hyper-heuristic approach to aircraft structural design optimization. Structural and Multidisciplinary Optimization. 2013;**48**:807-819. DOI: 10.1007/s00158-013-0928-3

[52] Xin J, Li S, Sheng J, Zhang Y, Cui Y. Application of improved particle swarm optimization for navigation of unmanned surface vehicles. Sensors. 2019;**19**:3096. DOI: 10.3390/s19143096

[53] Kaur J, Saxena P, Gaur P. Genetic algorithm based speed control of hybrid electric vehicle. In: 2013 Sixth International Conference on Contemporary Computing (IC3), Noida, India. 2013. pp. 65-69. DOI: 10.1109/IC3.2013.6612163

[54] Amertet Finecomes S, Gebre F, Mesene A, Abebaw S. Optimization of automobile active suspension system using minimal order. International Journal of Electrical and Computer Engineering (IJECE). 2022;**12**(3):2378-2392

[55] Liu Y, Qin Z, Liu J. An improved genetic algorithm for the granularity-based split vehicle routing problem with simultaneous delivery and pickup. Mathematics. 2023;**11**:3328

[56] Zhong H, Xu T, Yang J, Sun M, Gao F. Optimization design of automotive body

stiffness using a boundary hybrid genetic algorithm. Machines. 2022;**10**:1171. DOI: 10.3390/machines10121171

[57] Vinod Chandra SS, Anand H. Nature inspired meta heuristic algorithms for optimization problems. Computing. 2022;**104**:251-269. DOI: 10.1007/s00607-021-00955-5

[58] Ghorbany M, Ebrahimi-Nejad S, Mollajafari M. Global-guidance chaotic multi-objective particle swarm optimization method for pneumatic suspension handling and ride quality enhancement on the basis of a thermodynamic model of a full vehicle. Proceedings of the Institution of Mechanical Engineers, Part D: Journal of Automobile Engineering. 2023;**237**(14):3334-3352. DOI: 10.1177/09544070221148287

[59] Escoto M, Guerrero A, Ghorbani E, Juan AA. Optimization challenges in vehicle-to-grid (V2G) systems and artificial intelligence solving methods. Applied Sciences. 2024;**14**(12):5211. DOI: 10.3390/app14125211

[60] Gstalter E, Assou S, Tourbier Y, et al. Toward new methods for optimization study in automotive industry including recent reduction techniques. Advanced Modeling and Simulation in Engineering Sciences. 2020;7:17. DOI: 10.1186/s40323-020-00151-8

[61] Koopman P, Wagner M. Autonomous vehicle safety: An interdisciplinary challenge. IEEE Intelligent Transportation Systems Magazine. 2017;**9**(1):90-96. DOI: 10.1109/MITS.2016.2583491

[62] Gawron JH, Keoleian GA, De Kleine RD, Wallington TJ, Kim HC. Life cycle assessment of connected and automated vehicles: Sensing and computing subsystem and vehicle level effects. Environmental Science & Technology. 2018;**52**(5):3249-3256

[63] Frey CB, Osborne MA. The future of employment: How susceptible are jobs to computerisation? Technological Forecasting and Social Change. 2017;**114**:254-280

[64] Gurney JK. Crashing into the unknown: An examination of crash-optimization algorithms through the two lanes of ethics and law. Albany Law Review. 2017;**79**(1):183-267